About the Author

M. H. DUNLOP holds a Ph.D. in American literature from George Washington University. She teaches graduate and undergraduate courses in nineteenth-century American literature and culture at Iowa State University and has published widely on those topics. She is the author of *Sixty Miles from Contentment: Traveling the Nineteenth-Century American Interior*. She lives in Ames, Iowa.

Praise for
GILDED CITY

"Dr. Dunlop's picture of New York City at the close of the Gilded Age, a study in arrogant wealth, crude tastes, coarse vice, and cynical smugness, is a remarkable piece of research and writing, with Stanford White as the embodied spirit of the place—excellent reading."

—Jacques Barzun, author of *From Dawn to Decadence*

"The underworld of grifters and con artists, the extraordinary excesses of the nouveau riche, the barbarism of the so-called genteel society, all strut before us in this rich history of New York City's dark side."

—Lorraine B. Diehl, author of
The Late, Great Pennsylvania Station

"[Dunlop] has made shrewd use of newspaper articles and of contemporary foreign accounts of the American scene. . . . [A]n astute account . . . highly rewarding. . . . Dunlop reveals how many of our current obsessions arose from the passions of the recent past."

—*New York Times Book Review*

"The novels of James and Wharton had their share of scandals, but they have nothing on this lively cultural history which . . . crackles with debauchery and dissipation. . . . Dunlop furnishes not only vivid narrative but also historical context . . . and marvelous digressions."

—*The New Yorker*

"Dunlop has . . . brought back to life the great, forgotten scandals and sensations of 1890s New York. . . . Thoroughly entertaining . . . [offers] some sobering perspective on the excesses of our own cash-coated days."   —*New York Observer*

"Informative, interesting, and perceptive. . . . [H]er points about consumption and competition among the newly rich are well taken."

—Jonathan Yardley, *Washington Post Book World*

"[S]plendidly described. . . . Dunlop is a gimlet-eyed observer, precise and meticulous in her descriptions, witty and remorseless in her interpretations." —*Los Angeles Times Book Review*

"Dunlop abundantly fulfills her promise to lead readers on a peeping tour into those swollen houses [of the rich], dazzling our eyes with those jewels and gizmos while shocking us with tales of greedy pranks among the gadget-happy." —*Newsday*

"Dunlop [writes] with verve and clarity. . . . Though she takes many entertaining detours along the way, Dunlop is on a deadly serious mission here, intent on plumbing the depths of big-city greed, lust, and cruelty." —*New York Post*

"Dunlop is a keen observer and analyst of urban phenomena."

—*New York Daily News*

GILDED CITY

Scandal and Sensation in Turn-of-the-Century New York

M. H. DUNLOP

Perennial

An Imprint of HarperCollins*Publishers*

For Donald, Amy, Meg, and Ray

A hardcover edition of this book was published in 2000 by William Morrow, an imprint of HarperCollins Publishers.

GILDED CITY. Copyright © 2000 by M. H. Dunlop. All rights reserved. Printed in the United States of America. No part of this book may be used or reproduced in any manner whatsoever without written permission except in the case of brief quotations embodied in critical articles and reviews. For information address HarperCollins Publishers Inc., 10 East 53rd Street, New York, NY 10022.

HarperCollins books may be purchased for educational, business, or sales promotional use. For information please write: Special Markets Department, HarperCollins Publishers Inc., 10 East 53rd Street, New York, NY 10022.

First Perennial edition published 2001.

Designed by Jessica Shatan

Library of Congress Cataloging-in-Publication Data is available.

ISBN 0-06-093772-6

01 02 03 04 05   WB/RRD   10 9 8 7 6 5 4 3 2 1

CONTENTS

ACKNOWLEDGMENTS

I WOULD LIKE TO render thanks to Tom Kent, chair of the English department at Iowa State University, for his aid and interest; to Richard Major for his special historical expertise; to my friends and colleagues Kris Fresonke and Susan Conrad for their continual enthusiasm and helpfulness; and to Ed Goedeken and Rita Marinko of the Parks Library for their expert assistance. Most especially I thank my agent, John Ware, for his excellent advice and his untiring efforts on behalf of me and my book, and my editor, Toni Sciarra, for her cogent suggestions, her encouragement, and her wonderful high spirits. It is impossible to express sufficiently my thanks to my mother, great reader that she is; to my father, great man that he is; and to my husband Donald, who did everything he could and always more than I asked.

ON A JULY DAY IN 1906, Giulia Morosini invited the *New York Herald*'s society reporter to a private showing of the $2 million wardrobe she had readied for the upcoming social season. Morosini was the daughter of the banker Giovanni Morosini, a Venetian immigrant who had risen from a clerkship to a position as Jay Gould's right-hand man; he "permitted her," as she put it, "to indulge a cultivated taste or an aesthetic idea without concern for expense." Morosini herself was not only "famous for her wonderful gowns" but was also admired for having perfectly achieved the stunning pigeon-breasted and wasp-waisted body shape of the moment. For the *Herald* reporter, Morosini explicated, one piece at a time, the dresses, lingerie, and coats that her maid silently held up for viewing. The "famous gowns" made to her own designs in silk, chiffon, satin, and lace were both fragile and heavy, hand-embroidered, beaded, ribboned, and encrusted with diamonds. Because, in her words, "today we have no simple frocks for the laundry," she expected to wear most of the gowns but once. Morosini took care to point out diamond buttons and diamond horseshoe buckles as her signature touches and she put a price on each and every item. "One must pay," she announced, "for the name of an artist in gowns quite the same as in painting," and she summed up the overall effect of her bejeweled wardrobe as *artistic*.

Under so ferocious an onslaught of display, the *Herald* reporter apparently blinked, or winced, only when the maid held up Morosini's final clothing triumph: a coat of white unborn baby lamb lined with ermine. The reporter's reaction elicited from Morosini a dominant platitude of the time: "Money spent in this way is not lost," she asserted, "for if the dressmakers and milliners and shoemakers had no demand for their work the wheels of progress would necessarily be hampered." Then, taking care to cover any remaining weak points in her display and gripped by a longing to drape over her rapacity something of the beautiful, she tucked her display behind the late nineteenth century's greatest protective shield. The goods, she said, were *artistic*.

Giulia Morosini was right to display her wardrobe; Americans wanted to see other Americans' goods. To show the goods, however, was risky. The very act of display suggested that one was courting audience approval and admiration, but the audience was just as likely to answer an outrageous or extreme display with scorn or disgust. Moreover, even though the audience very likely wished to know the prices of the items on display, an individual showoff risked looking mighty crude when he or she recited the price of each item on show. There was further risk entailed in a public attempt to justify staggering personal expenditure. In the end, even if Giulia Morosini was, in a sense, right to display her wardrobe to America, she turned out to be wrong about everything else.

When Morosini claimed that production of her extravagant wardrobe had made a genuine contribution to the social welfare and had constituted a real effort to fight unemployment, she seriously mistook her relation to the society she lived in. By 1906, the social-welfare justification for wild consumption—always a dubious claim but one invoked by the rich for more than two decades—had lost its legs. Giulia Morosini was nobody's benefactor if she could wrap herself only in the tattered fantasy of a world wherein the eternally poor labored to render service to the perpetually rich. Her invocation of that tired notion did no more than confirm her insensitivity to and ignorance of the lives of other beings, both human and animal. The most notable effect of her wardrobe appeared not in trickle-down profits but in her own lost awareness of others. The arrogant instructional tone she took toward the *Herald* reporter reinforced her misperception of her real relation to society, a relation of which she was necessarily and determinedly unaware. Racks of jeweled clothing stood between Giulia Morosini and the rest of America.

In social terms, Giulia Morosini was neither a benefactor nor a taste leader but an American who had earned nothing, invented nothing, made nothing, and thought nothing. As a very rich woman, she had plunked herself down on one of the very few chairs that society made available to her. And although that position was, in the parlance of the time, a "high chair," it was also an uncomfortable one. In it, Morosini had become, literally, a being surrounded, shaped, and consumed by her expenditures. Not only had she radically deformed her body to suit the demands of her wardrobe, but she had also allowed her mental energies to be entirely intercepted by her complicated goods. The goods, in their turn, had consumed her without giving her anything to think about. In struggling to see herself as the creative spirit behind her goods, Morosini was further mistaken. She appeared at one of those moments in the history of spending when goods, in order to remain out of reach for the average person, so increase their complexity as to make immense physical and psychological demands on the few who own them. At those moments of extreme pressure from too much money and a wide availability of goods, Americans can conceive only of making life and its goods more complex and more demanding of themselves and their time, and consumers who long to see themselves as special are forced to abandon any lurking American hope of simplicity. Finally, Morosini failed in her desperate lunge for shelter in the *artistic*. Every detail in the *Herald* article said her clothes were not *artistic* but gaudily difficult, and thus Morosini herself was no artist. She was a finicky and imperious shopper who had made her own body the carrier of as many items from the stock list of desirable goods as it could bear. Encumbered, in fact, by many more goods than ideas, she signaled no personal awareness of either the weight of her consumer burden or how, staggering under it, she might appear to others.

Diamonds, animal hide and fur, along with orchids, foie gras, big art, elaborate décor, deluxe gizmos, showy vehicles, and colossal houses made up the hot list of desirable stuff at the turn of the last century. This unsurprising list signals a standard Western vision of luxury to which America's only distinctive contribution was the gizmos. In

America, the luxury vision blooms fully during those special moments when the country feels unthreatened by outside forces and when a stratum of suddenly moneyed and suddenly visible Americans throws itself into spending as fast as possible the unexpected millions that have fallen into their hands. These Americans neither invent the list of desirable goods nor make any additions to it; they discover it complete, hanging in the thin air of the past, awaiting revivification in the hands of big new money. The list itself is not neutral and not benign but active and dangerous; the items on it have their own life histories, their own behavior patterns, and their own counts of historic casualties. Their sheer power as goods is enhanced if their latest round of adherents is helplessly unaware of how these goods have previously behaved in the past. Moreover, the moments at which gigantic houses and multiple diamonds and troubling furs reappear are noisy and jarring times, moments when Americans claim, as they did at the turn of the old century, that their lives are undergoing breathtaking increases in pace, are under assault from staggeringly rapid change, are so complex and fast moving that no one caught up in them can afford to pause, to think, or to analyze.

It is at such moments that the objects, so to speak, can move forward into action. They are not, after all, mere insentient things but powerfully standardized stuff equipped with standardized behaviors, with old established story lines of their own and with risks that their owners perhaps cannot see but will be made to experience.

Between 1880 and 1910, in a period dominated by big new money and by the perception that the pace of American life was drastically on the increase, a time when city dwellers forced themselves to go "the pace that kills," the desirable goods of the time also swung into motion. Jewels enlarged and multiplied, overwhelmed their settings, and thickened into weighty clusters. Houses swelled to gigantic proportions, disconnected awkwardly from their surround, and then required their owners to conceive of some way to live in the cavernous spaces they had commandeered as a sign of their power. Vehicles enlarged toward the unmanageable or compressed down to the privately tiny. Deluxe and desirable gizmos—always promoted as useful and time-saving—demanded rapid acquisition and equally rapid replacement, since the most successful technological doodads lost their special luster as soon as they became cheap and widely available. As a group, these standard luxury goods teased and bedeviled and challenged their owners, and along the way they took charge of many lives into which they entered. Their effect was blinding; for a time, the rich and many others in their ambit lost sight of connection to anything but money and goods.

In the grip of their stuff and equipped only with the skimpy aesthetic of *more* and *larger,* turn-of-the-century Americans wielding the power of money took to experimenting with human resources and with what other Americans could be made to do in their service. When these experiments burst into public view, they produced sensations felt across the country and even around the globe. Notable late-nineteenth-century party givers set out to discover how many Americans could be made, on demand, to dress up in pink satin and gold lace, how many men would shave off a carefully cultivated and fashionable mustache so as to more closely resemble Louis XVI, how many men would display their legs in the service of a general royal fantasy, and how many hired detectives would powder their hair if offered ten dollars apiece to do it. Rich and not so rich "bachelors" and "men-about-town" set out to discover how many girls—at fifteen dollars apiece—would drop their undergarments in front of a group of men. If the social rich learned those things—and they did—they also learned, when their high jinks came into public view, that their fellow Americans were watching them and wanted to examine not only their goods but their errors, errors which were as standard as their goods. The citizenry's desire to remark, publicly and at length, on their missteps inevitably struck the rich as early warnings of impending social chaos.

The rich and those mesmerized by the rich suffered equally under the lonely suspicion—Giulia Morosini's suspicion—that the best use of one's personal resources and of the resources constituted by other human beings lay in the enhancement and positioning of the self. Persons low on resources could participate in the suspicions of the rich only by creating fantasies of access to resources. Then they worked to transform the fantasy into reality: they dressed and acted and bought as if they had money. Some worked the fantasy of wealth by amassing so much stuff that others might necessarily assume them to be rich; when creditors and bill collectors refused to support the fantasy, these fantasists underwent sensational and instantaneous social wipeout. Wild overspenders further learned, in those terrible moments of discovery, that other Americans would not take their excess goods off their hands at a decent price. Those goods were, after all, the remains of personal disaster, and as such they stank of failure. While moneyless spenders risked personal chaos, other fantasists—those who wanted not tons of stuff but only the consumer's thrill of placing orders for it—created real social disruption when they popped from one office to the next, scattered phony business cards about, placed orders everywhere for heaps of goods they intended neither to pay for nor to receive, and then vanished.

As the goods seized the full attention of their owners, other Americans—unless they had matching stockpiles and thus served as accomplices—became more remote and increasingly incomprehensible. The rich mistakenly expected the rest of society to quietly put up with them while they tried to find their way, but they continually stumbled into the path of those troublesome and mocking Americans—of whom there are always a few available—who had ideas and who refused to just shut up and shop. At the turn of the last century, such objectors to the dominant trend were regularly classified as "crazy"— that is, as persons whose ideas and actions had neither meaning nor importance. When it proved impossible to have every social critic declared officially insane, the very rich retreated to the close company of their own. And when, in their own world, they wearied of seeing only persons like themselves who owned the same things they owned, when they suffered a boredom too comfortable to qualify as real suffering, they moved in new and risky directions. They launched secret lives from which they excluded those nearest to them, or they turned, in that heyday of American addiction, to alcohol and drugs, or they sought the thrill of watching other beings suffer in ways that were closed to them. They cultivated insensitivity to the suffering and discomfort of others: they went on midnight slumming tours and sneaked peeks at the dirty feet of the unimaginably poor, they studied the faces of children contemplating toys they would never have a chance to touch, and, once a year, they charitably set a decent meal in front of persons certified as not having consumed such a good meal in the past calendar year.

Next came the animals' turn to be caught up in what seemed a general American effort to lay claim to—if not to exhaust—every possible available resource. What they did to animals was not historically new but it was, at the turn of the century, newly intense. Animals, lacking even the low resistance of the human world, were made to yield up their decorative and entertaining possibilities, to give themselves over to human display, and to serve dead or alive as décor. Some died horribly when they failed to understand what humans wanted of them or failed to gratify the desires of their human keepers. It was a sad day for the animals when their personal coverings entered the category of luxury and became objects of desire. There is no limit to what can be done to an animal once it has been aestheticized or turned into an expression of do-as-I-please dominance, and it is impossible to do more than guess at how many ewes and prenatal lambs were sacrificed to Giulia Morosini's coat.

Consequences, of course, were to follow. Underpaid and mistreated servants decamped as soon as they could and went off to lead their own lives far from the inexperienced and imperious employers whose manners failed to rise to the level of their situations. Colossal houses proved unlivable and, when no second owner appeared willing to take on the burden of another's taste choices, they fell to the wrecker's ball. When the houses fell, the elaborate goods and the big art were scattered. In the technology-crazed America of the late nineteenth century, new gizmos rapidly lost their cachet and left rich early adopters feeling like nothing more than part of a well-equipped crowd. And when tastes and values changed, those who had not failed by losing could not win by winning: sums of money and heaps of goods that had seemed to promise something unlimited, some prize of new freedom and power, turned out to withhold that promise. Meanwhile, ordinary Americans lost interest; they increasingly shifted their attention away from the social rich and toward entertainers trained to make a more pleasing display of themselves than the rich had ever figured out how to do.

These are big-city moments in a period that saw the city come to dominance in America. Some of the city persons who appeared during these moments are still recognizable names; others once famous are now forgotten; and some were never anything other than obscure. No matter: at the center of each of these moments is a clash of desires and ideas about having, seeing, touching, displaying, performing, and killing. Contemporaneous voices arguing over who could look, who—if anyone—could touch, and what money could buy emerge here from the two major pools of material used in the writing of this book: first, approximately one hundred books of travel, description, observation, and analysis, recorded on the spot and at the time by both locals and visitors from abroad; and second, the city newspapers that furnished a daily record of the ongoing conversation that constitutes American culture.

Although a few late-nineteenth-century books of travel and description have had the staying power to be reprinted, many were lost, and the survivors—printed as they were on the cheap acidic paper used in postbellum America—were turning to powder even as I read them. Some were sole surviving copies and some had never had their pages cut. I am grateful to the repositories that loaned out these books for my use; after all, why let a book die on a shelf when it could just as well die in the hands of one last interested reader? Aside from their terminal physical condition, the texts themselves carry great immediacy. Their writers were astonished by America's big cities, anchored on the grid—which every foreign visitor deeply admired—and yet fluid in process, changing under the eyes of the observer no matter how short his or her visit. The books are filled with argument and complaint and dark predictions, but they are also filled with admiration for big-city pace, style, and variety, and they are shot through with deep appreciation for America's dedication to the technologies of human comfort.

America's big-city newspapers—described and analyzed at length by every foreign visitor in the 1880s and 1890s—constitute the other great body of source material for this book. At the turn of the last century, the great city dailies were everyone's indispensable reading, the very framework on which a day in the city was hung. Reporters patrolled the city streets around the clock, uncovered the unannounced and the unthinkable, and pushed thousands of persons otherwise lost to history onto the city stage for their moment. The newspapers of the old century are not about their time and they do not now refer to some reality that exists behind them; they *are* that reality and they *are* their time. Furthermore, to immerse oneself in the newspapers of the past is not to find a reductively "quaint and simple" past, safely ended and cut off from the present. Instead it is to find a fresh mix of the familiar and the unfamiliar, a powerful shot of discomfort, and a deep unease.

Daily big-city newspapers in the late nineteenth century lived by narrative and by unquestioned belief in the special abilities of the trained eye. Happily for them, no false and mechanistic *objectivity* interfered with their belief in the sharp eye, the immediate impression, and the insolent question. The dailies snapped and popped, they offered up a confusing and uproarious world unfiltered by rumination, and they thereby saved the world from the deadening middle-class gentility of the periodicals. Foreign visitors were lost in admiration of America's big-city newspapers and the range of their influence. Americans, they noted, were "voracious newspaper readers" and newspaper reading was a "fixed national habit." The newspapers offered "densely packed columns in which one can obtain, better and more easily than by any other means, an idea of the customs, manners, and civilization of the people," and in those columns there was "such animation, such activity, such fever, such push as old Europe has never dreamed of." Not only did the daily paper "fairly quicken the excitements of everyday life," but the Sunday edition "takes the place of church with non-churchgoers, and in order to meet a want, it usually supplies a sermon." For some observers, however, the latter assertion had a dark side: "Nothing is holy in the eyes of the American press; it respects nothing."

Eight dailies make contributions large and small to the material of this book: the *Chicago Tribune,* the *St. Louis Post-Dispatch,* the *New York Tribune,* the *New York Sun,* the *New York Journal,* the *New York Telegram,* and the *New York Herald.* The *Herald* is its bedrock. Throughout my life as a devotée of newspaper reading, I have never encountered a better newspaper than the *Herald* was in the 1890s; its readers at the time placed it "at the front rank of the American press" and judged it to be "an extraordinary paper—a paper that is the most seen and the best informed in the entire world. The *Herald,* in short, is a universal encyclopedia of the day."

The *Herald* spent lavishly on producing broad international coverage not only of politics but also of art, style, and taste. On its American front, the *Herald* knew so much about the filthy rich that it could needle them, while less well informed newspapers could do no more than gasp and moralize. On the other hand, the *Herald* was so sharply conscious of upper/lower class jostlings and tensions that its interest in middle-class life was low. Rural and small-town America—a population invariably referred to as "the country cousins"—it relegated to joke material, as in the following attempted-suicide headline of April 9, 1893.

NEW YORK GAS A FAILURE
A Hayseed Blew It Out and Went to Bed and Didn't Die

All of which, of course, makes the *Herald* the ideal city sheet. The *Herald* did not promote city life and it never took an educational or therapeutic view of city experience; the *Herald* simply believed in cities. In the pages of the *Herald,* the known universe was a series of great cities strung across the globe, and city life was a highly charged drama, chaotic, inconclusive, funny, and fiercely exciting. Each of the players in that drama, whether human or animal, the *Herald* found worthy of notice—at least for the moment—and, in the parlance of the time, it served up each of them "piping hot."

The "piping hot" category was also the category of the uncontrollable. Hotly desired goods were troublesome goods—hard to control, quick to reproduce, uncertain in value, and perhaps even determined to misbehave. Terrible personal and social moments occurred when the goods—material, human, and animal—turned on their putative temporary owners. Then parties went out of control, titled persons behaved rudely, barndoor-size paintings drew snickers, bric-à-brac drove the houseguests mad, sale goers refused to bid on the second-hand décor, exotic dancers threatened to tell all, and animals refused to be "good." Americans publicly embarrassed by their goods and unable to gain control over them might exercise one of several available options, none guaranteed to succeed. They might demand silence, though they never got it; they might attempt to conceal the objects or persons they had failed to control; they might ship everything to storage, resale shops, and auction rooms; they might pull the shades, leave town, and return only when their sensational behavior had been replaced in the public eye by someone else's sensational behavior. Or they might exercise the most drastic of options—and often the most sensational too—and move to destroy those troublesome things.

GILDED
CITY

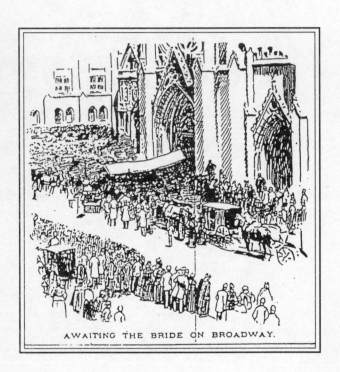

AWAITING THE BRIDE ON BROADWAY.

CHAPTER 1

Diamonds on the Mind

To the Waldorf Hotel, January 31, 1897. Dear Gentlemen—Let me say in a very few words that there is a great scheme being contemplated of dynamite destruction in your hotel on the night of the Bradley Martin affair, and by a person whom you least expect. I dare not say too much now as I think I am being watched by the gang, but keep your eyes wide open till you hear from me again. I will not dare to disclose my name at this moment. Bye-bye. In earnest.

AT ONE IN THE AFTERNOON of Tuesday, April 18, 1893, at Grace Church in Manhattan, sixteen-year-old Cornelia Martin was married to twenty-five-year-old William George Robert, fourth Earl of Craven and Viscount of Uffington. She was the only daughter of Bradley Martin and Cornelia Sherman Martin, and he was a heavily tattooed English country boy, rumored to be impoverished, who appeared at the church with his trousers rolled up above his boots. His contribution to the union was his title, while the Bradley Martins, Americans living in Scotland, contributed to his future well-being the equivalent of eleven million dollars. Perhaps in an effort to raise Cornelia Martin above the mere countess that she would become, the Bradley Martins described the wedding presents as "fit for a royal princess." With a show of reluctance, they released to the New York newspapers a short list of gifts said to "exceed anything ever given at a wedding in this country," including "a diamond tiara, a bracelet set with alternate diamonds and sapphires, a collarette of superb Indian stones, old mine diamonds, set in silver, a vinaigrette of Louis XV style set at the top with a precious topaz surrounded by diamonds, a globular watch encrusted in diamonds attached to a diamond coronet, a dagger of pearls and diamonds, a ring set with three solitaire diamonds, a sapphire and diamond dagger, a hat pin set with diamonds and sapphires, and a brooch composed of a wreath of diamonds."[1]

The Bradley Martins experienced life as a series of dollar amounts and a stream of goods, each with its price tag. The experience was not entirely satisfactory. Especially in America, land of property-tax assessors and social critics, life for the Bradley Martins lacked the comfort of entitlement. In their yearning for settled and predictable public homage, the Bradley Martins sought to anglicize and royalize themselves while masking that effort as entertainment for others. In their struggle, they had the advantage of being exquisitely attuned to the great consumer icons of their time, to the visible and desirable objects of agreed-upon value: they knew what to select from the enormous repertoire of urban consumer goods, and they knew how to transform clusters of those goods into a social event. Moreover, they knew how to elaborate the pleasures of consuming: they displayed more of the desirable stuff than other persons did, they persuaded their guests to give them the same icons they displayed as hosts, they teased the multitudes into wishing for a look at their goods, and they exacted homage to themselves in the form of publicity. They had no psychology; they were, to steal a phrase from Henry James, persons of "bottomless superficiality." It is as Americans that they matter. Even when they moved to England and hinted that their name might be hyphenated as Bradley-Martin, they remained Americans, and they were never more American than when, under the pressure of American laughter, they backed away from hyphenation and claimed that it had been a newspaper error and none of their doing.

At the top of Mrs. Bradley Martin's list were jewels, objects whose desirability was unquestioned and accepted across American society. Mrs. Bradley Martin owned jewels in "every conceivable form," and indeed had amazed Scottish locals when she turned up at county events in Inverness-shire wearing her tiara of solitaire diamonds, or her Marie Antoinette rubies, or her diamond sun. Because local reaction to her in Inverness-shire had been mixed, she had sought out rich expatriates like herself and had taken to tossing parties at which she could be seen by no fewer than forty people who really understood diamonds. Across the class levels of the Bradley Martins' circle, everyone was called upon to affirm her values, including the servants: the butler, two footmen, and Mrs. and Miss Martin's maids together gave Cornelia Martin a bracelet set with "a beautiful pearl surrounded by diamonds," piously pronounced by the Bradley Martins to be "one of the most valued wedding presents of all." The groom gave diamond lace pins and scarf pins to the members of the wedding party, and quite a few other jewels changed hands when numerous guests and onlookers were robbed of their jewelry in the crush at Grace Church.

New Yorkers had been primed for Cornelia Martin's wedding by intense daily newspaper coverage of gifts, titles, and floral decorations; publicity engineered by the Bradley Martins had teased the locals into wanting to see not only the bridal couple but also the twelve hundred fashionable guests expected out of the three thousand invited. Three hours before the event a crowd began to gather for the show. Broadway strollers joined the throng, and street cleaners and bootblacks abandoned their work and were swallowed up in the crowd. The stretch of Broadway between Tenth Street and Thirteenth Street was transformed into a human mass, and in each window of the St. Denis Hotel opposite Grace Church stood persons equipped with binoculars and opera glasses. Because no accommodation had been made for the uninvited onlookers, they created their own viewing protocols. As each carriage approached from the north, a thick and jostling crowd halted it and settled six deep around it while they examined its contents, discussed with each other the looks and the lace of its occupants, and assessed the quality of their diamonds. The uninvited freely questioned carriage occupants about the bride's millions and the heraldic tattoos reported to cover the chest, back, and biceps of the groom.[2]

As carriages neared the church doors at Tenth and Broadway, their occupants confronted an even thicker crowd of persons who were not only ready to get rough but were also "rather inclined to be critical." Meanwhile, the police began to shove people about freely, and tramcar drivers tired of whistling for space drove deliberately into the crowds. Wagons heaped with merchandise, a funeral cortège, a police patrol, a fire engine, and an ambulance were sucked into the "maelstrom," and a horse that had broken loose at Great Jones Street dashed up Broadway at a gallop and was hit by a truck directly in front of Grace Church. Some of those on foot were crushed in the melee, and occasional cries of pain shot up from the dense interior of the packed crowd. Panic threatened. The *Tribune* noted that "there was no outburst of applause even when the bridal party, smothered in great white bouquets and streaming with white ribbons, drove up, and a vision of white and nodding ostrich plumes swept into the church."

Inside Grace Church waited both the invited and the uninvited. Many invited guests who arrived after twelve had difficulty squeezing into the already packed church; the uninvited had made up their own invitation cards to display at the church door, and the police waved them in without examining the cards. Among fashionable men, attending the ceremony itself was considered "not exactly the thing to do," so the congregation consisted of forty women for every one man. The Bradley Martins had no connection with organized religion: in order to be married in Grace Church, their daughter had been hurriedly confirmed the week before the wedding. Furthermore, they assigned no particular value to music: the regular church organist, Samuel P. Warren, played standard bits from *Tannhäuser* and *Aïda* while the church bells offered "Believe Me If All Those Endearing Young Charms." Flowers, however, they cared about; flowers could be counted, graded, measured, and differentiated. Guests walked through a "vista of great Bermuda lilies" trimming each pew. In the chancel, palms stacked to a height of forty feet formed a backdrop for banks of daisies, hydrangeas, lilies, roses, azaleas, and rhododendrons reaching to a height of thirty feet. The four chancel columns were festooned with spring flowers, the chancel rail was buried under mounds of lilies of the valley, and the steps leading to the altar were carpeted with loose rosebuds and lilies. Each of the Bradley Martins' liveried servants wore a huge white wedding flower, the bride herself carried a bouquet of white orchids, and the bridesmaids toted bundles of white lilacs and Mabel Morrison roses.

When the bride, looking "awe-stricken over her voluminous satin gown and fine lace and jewels," and her father stepped into the aisle, one short guest, determined to see the whole show, clambered up onto the seat of her pew, and then the guest whose prospect the climber had blocked did likewise, and within moments five hundred wedding guests were, according to the *New York Herald,* "not only standing on the pew seats but actually balancing themselves on the backs of the pews." The *Times* reporter, with a view from the side of the church, watched guests stack kneeling cushions on the pews and then attempt to balance themselves atop the heaps of cushions. The guests chatted throughout the ceremony, just as they were accustomed to do at the opera: they discussed the dollar sum that the Bradley Martins had settled on the Earl, and they passed a rumor that a certain woman, present in Grace Church but unidentifiable, "took more than a passing interest in the wedding." They criticized the floral decorations from both artistic and financial points of view, and they wondered why the Earl's aunts had favored cutting short the engagement period and rushing the wedding.

The big ruckus, however, began when a sharp-eyed someone in the crowd outside the church discovered an open door on the north side of the church; apparently Mr. Partridge, the sexton, had anticipated the end of the ceremony by several minutes. *En masse* the uninvited crowd surged toward the open door. Just as Henry Codman Potter, Episcopal Bishop of New York,[3] began to pronounce his benediction on the couple, the crowd burst into the church with a great triumphal shout punctuated by shrieks of pain from some who were trampled to the floor and others who were lifted up above the crowd and tossed into pews occupied by the Bradley Martins' guests. Pews splintered and gowns ripped. The police outside managed to shut the door, thus effectively trapping a good-sized segment of the invading crowd inside the church. When the invited guests departed, the invaders "began to take away souvenirs of the occasion, and men and women stripped the altar of its beauty by tearing down vines, cutting roses, and breaking off entire stems of Easter lilies." The uninvited liked flowers too, and not until the police fought their way into the church and threatened them with arrest did they stop helping themselves to the décor.

When the bridal couple departed, a screaming segment of the crowd that had not got into the church chased the bridal equipage up Broadway to Fourteenth Street in a final attempt to have a look at the occupants. "No such spectacle," announced the *Herald*, "has ever been seen before in New York, and may it be many decades before its like is repeated." A day later, everyone involved wanted to gloss over the disorderly scene. The newspapers claimed that "good humor"—a powerful American value—had pervaded the socially tense Grace Church scene, and the Bradley Martins sighed to newspaper reporters that they regretted "only that the crowd was so great, but that, after all, was an indirect compliment to so brilliant an occasion." They did not, however, see it as a compliment when that very same night they were robbed of not only the thirteen nicest specimens in Bradley Martin's watch collection but also most of the silver that Mrs. Bradley Martin had received at her own wedding in 1869. Although the burglar did not get a look at the wedding gifts, which had been locked in a safe, he or she did perform a series of eloquent and stylized social gestures: the intruder used a glazier's diamond to slice through a window, took eleven complete silver place settings, arranged the remaining twelfth setting on the floor, and departed through the front door. In the general confusion over who was entitled to do what, the *Tribune* said the thief was just like the persons who had stripped Grace Church of its flowers after the wedding; the *Herald*, on the other hand, said that the flower-strippers were just exactly as crude as the pew-climbers. Nonetheless, the style of the burglary sent a message, a message of contempt rivaling that delivered by the Earl of Craven when he appeared in Grace Church with his trousers rolled up as if in preparation for wading into a hog wallow, a message to the Bradley Martins that they could eat off the floor from now on. Wit was not lacking, but "good humor" seemed to be faltering.

The Bradley Martins were, mostly by their own design, lightning rods for social notice of several kinds. They were not personally interesting in any way, but there was, according to the *Herald*, "a fashion that people have fallen into of celebrating and criticizing their slightest movements." Their central story line in 1893 involved directly trading their pathetically young and inexperienced daughter for a heap of diamonds, three thousand gifts that they valued at three million dollars, and an alliance with a title whose emotional connections lay elsewhere. Covertly this story line was fully understood, but overtly it was only hinted at. Cornelia Martin herself had no story. No one bothered to invent a love story about her marriage, and no one writing up the wedding attached any personal sentiments to the bridal couple or even suggested that they had any interest in each other. In truth, the union engineered by the Bradley Martins quite resembled the exchanges of money, status, youth, and looks proposed to strangers every day in the newspapers' personals columns.

> A thorough gentleman (35) desires acquaintance refined lady, about 30, dark blonde; substantial figure; residing convenient Christopher Street ferry; object, matrimony.

> A refined, educated gentleman, of pleasing appearance, aged 48, highly honorable, strict business integrity, would like to correspond with a refined lady, aged from 30 to 45, who has $100,000 to $150,000 or income of $5000 to $10,000; object, marriage.

> A refined young woman of 19 wishes to meet well bred man who can appreciate and afford the luxury of a well groomed companion; object, matrimony.

> A young gentleman of noble, titled family, with every manly accomplishment, and most brilliant prospects, desires to meet a lady of means; object, matrimony.[4]

The personals were specific about money, weight, and place of residence, but silent on the subjects of personal attraction and human compatibility. Those few wedding guests who could be persuaded to comment on the Martin-Craven wedding made no remarks about the match itself and offered no stories; they questioned only such matters of taste as the pink shirts worn by the ushers. American newspapers, displaying the long-standing American penchant for record-keeping, attempted to position Cornelia Martin as "the youngest Countess of Great Britain," but that status was too obviously temporary to be of much interest. The fourth Earl of Craven's story that his brilliantly colored heraldic tattoos were the product of a "secret acquaintance" with an old sailor was too obviously bunk. The only narratable story related to the wedding was a tale about money, a whopping lie, and a soiled dress.

On April 5, 1893, the Bradley Martin party disembarked at the Port of New York and the group left Mr. Bradley Martin alone to handle their 128 pieces of baggage. It was Cornelia Sherman Martin who directed spending on the jewels, costume events, flowers, and public situations that put her in the spotlight. Bradley Martin was a considerably dimmer figure, devoted to hunting and fishing but otherwise known only for his collection of watches. Although it is tempting to think that Bradley Martin collected watches because he bore a certain residual American preoccupation with time, it is likelier that he was imitating the watch collections amassed by such wildly acquisitive European titles as Ludwig II of Bavaria. On this occasion, however, Bradley Martin had been put in charge not of spending but of ensuring that as little money as possible vanished into situations that yielded neither goods nor amusement nor publicity. Like many others with both more and less to spend, he preferred to pay out as little as possible to servitors and assessors while sparing no expense on the great material icons of the time. So on April 5, while Surveyor of the Port George W. Lyon and his several assistants waited at the pier in expectation of viewing "considerable dutiable property," Bradley Martin played his part and declared nothing but a picture of his son and a barrel of oil that he valued at twenty-five dollars. He told the astonished Deputy Surveyor John Collins that the wedding trousseau had been left behind at the Martins' "Scotch castle" and, when asked about Cornelia Martin's wedding dress, he said that it was in one of the trunks but was old, considerably used, quite soiled, and thus not dutiable. He swore an oath to same. When the inspectors opened the trunk containing the wedding dress, they assessed it as new, but Bradley Martin swore a second oath that it was soiled and consequently nondutiable, and the inspectors found themselves forced to take him at his word. During the next few days, however, customhouse officials allowed a rumor to spread that a secret inquiry was under way, and that "if it is found that Mr. Bradley Martin's object was to evade paying duty, the wedding dress and other dutiable articles would be seized."

Nothing happened. The soiled-dress story line ended with the Bradley Martins taking some brief public umbrage and undoubtedly some extended private glee. From the Bradley Martins' point of view, after all, they had come to the United States to put on one of the shows that they launched at America every three or four years throughout the 1880s and 1890s. Why they should have to pay to import the show's principal costume was not within their understanding.[5]

The Bradley Martins had but one daughter and thus but one chance at a title in the family; they seized it, perhaps in the fear that another might not come along. In ceremonially handing over their little girl and their eleven million to the Earl of Craven, referred to in the newspapers as "their prize," they had not only put themselves on show but had also assembled a potent lineup of the big-city icons of desire: satin and velvet costumes, diamonds, cut flowers, and an "aristocrat" whose brass remained untarnished even though, to casual observers, he appeared to be no more than a slight, tattooed, oddly dressed fellow with a known "fondness for theatrical persons." On their next foray into New York society, in 1897, the Bradley Martins would expand their display to include twinkling incandescent lights, an even more extensive deployment of silk and velvet, and lavish use of the color red, thus playing to the full list of objects that together constituted a vision of the artistic, the beautiful, and the valuable. The end of the nineteenth century witnessed the most serious effort ever seen in America to force objects into expressiveness and to squeeze out of them the rewards they seemed to promise; no greater demand was ever made on materials. Late-nineteenth-century Americans believed in their goods. The greater the number of objects, the greater the promise.

In New York City at the close of the nineteenth century, there was more wealth in private hands, more stuff available to buy, more opportunity to get ahead, and more densely packed poverty than anywhere else on the face of the earth. It was the best stage on which to launch a tense drama of consumption and the best locale for testing the potential meaning of desirable goods. Ideas that have worked their way so deep into the American psyche as to become catchphrases, such as Thorstein Veblen's famous "conspicuous consumption," are inadequate to dealing with the complications of the period. Competition among the rich is, for example, too narrow and exclusionary a cause to explain the phenomena of consumption, and when patterns of ownership and display cross classes, the resulting circularity of desire is not understandable through a theory of emulation.[6]

Veblen drew a dividing line between needs and luxuries, but in the closing decades of the nineteenth century, big-city Americans did not. Desire converged upon specific objects even though resources to buy remained drastically divergent. Whether a city person displayed a single diamond or a chest covered with them, a single flower in the buttonhole or rooms banked with flowers, a velvet hat or a boudoir draped in velvet, they agreed on the value of the diamonds, flowers, and velvet they were displaying, and they further agreed on their suitability for both room décor and personal décor. Consumption might cross class boundaries, but it neither demolished nor altered those boundaries. Furthermore, to Americans living in a whirlwind of goods and desires, notions of "happiness" were irrelevant. Big-city America had its eyes on its own distinctiveness, on France, on money, and on goods.

The Bradley Martins were transfixed not only by jewels and flowers but by themselves. "In spite of his supposed indifference to public notices," crowed the *Herald,* "Mr. Bradley Martin has busied himself collecting all newspaper mentions of their affairs, and already has several ponderous volumes of clippings." Furthermore, the Bradley Martins were as deeply confused about what they were up to as were the audiences for their social performances. They were confused over their national identity: the Earl of Craven's relatives had wanted to hold the wedding in Great Britain, but the Bradley Martins suddenly felt like the Americans they had once been and insisted on a wedding in America; their daughter, however, had never lived in the United

States, had few acquaintances in New York City, and purportedly did not even know her own bridesmaids. Once in America, the Martins shifted cultural identity again and insisted on having an "English-style" wedding; its vaunted "Englishness" was unclear to Americans, and seemed to involve mostly the marching order of the bridal party and the fact that the wedding cake was adorned with the Martins' "coat of arms." What is clear about the Bradley Martins is their ability to transport everyone into what anthropologists call an object world, a space within which the foremost focus of every participant, whether guest or invader, was consumer goods.

On the day of her daughter's wedding, Mrs. Bradley Martin was "radiant in smiles and huge diamonds," and indeed the twilight of the nineteenth century was lit by diamonds. Diamonds were, for two decades after 1874, neither rare nor expensive: two million carats of diamonds mined in South Africa within a two-year span saturated the market, and the resulting crash in diamond prices—a per-carat drop of about 60 percent—gave diamonds an availability they had never had before. Included in the dazzlingly bejeweled population of late-nineteenth-century New York were the rich, the aspiring, the nearly poor, men and women, oldsters and infants. Urban Americans wore diamonds as rings, cufflinks, necklaces, tiaras, stomachers, earrings, hair ornaments, brooches, bracelets, watches, studs, glove buttons, dagger pins, belt buckles, shoe buckles, hatpins, pendants, sunbursts, tassels, and charms. Although the setting mattered far less than the jewel itself, diamond-encrusted daggers, sunbursts, lizards, and insects, especially bees, had a great vogue in the 1890s. "Very swell babies," observed the *Herald*, wore "gold buttons set with tiny diamonds." Altogether New Yorkers owned today's equivalent of fifteen billion dollars in precious stones.

Diamonds were no luxury; they were a necessity for the presentation of self, the first purchase made by anyone who had a success in city life, desired and worn and given and owned by anyone who played a part on the urban stage or aspired to do so. Diamonds were irrelevant to quiet middle-class evenings at home; they were public jewels, meant to be noticed, gaped at, admired, desired, lost and found, pawned and recovered, stolen. Anything having to do with jew-

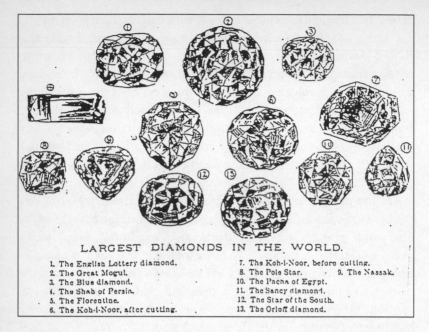

LARGEST DIAMONDS IN THE WORLD.

1. The English Lottery diamond.
2. The Great Mogul.
3. The Blue diamond.
4. The Shah of Persia.
5. The Florentine.
6. The Koh-i-Noor, after cutting.
7. The Koh-i-Noor, before cutting.
8. The Pole Star.   ·   9. The Nassak.
10. The Pacha of Egypt.
11. The Sancy diamond.
12. The Star of the South.
13. The Orloff diamond.

els was of interest. Newspapers offered their readers regular doses of diamond lore, diamond legends, and cursed-diamond stories, accompanied by artist's renderings of multijewel pieces and life-size drawings of "The Largest Diamonds in the World." Popular giftbooks carried such titles as *Lingua Gemmae* and *Jewel Mysteries I Have Known*. During Gounod's *Faust*, the favorite season-opener of the Metropolitan's boxholders, many operagoers stopped conversing only during Marguerite's "Jewel Song." Meanwhile jewel thieves were using the opera house as their workplace: wearing evening dress, ermine cloaks, and big diamonds, they concealed their intent by looking just like those from whom they intended to steal. Richard Wells, in *Manners, Culture and Dress of the Best American Society*, instructed his socially aspiring readers that "the diamond stands supreme among precious stones. The brightest among gems, it outshines all others. It was with diamonds that the angels tempted the daughters of men; with diamonds that Mephistopheles caused Marguerite to be tempted by Faust. Indeed, the fatal light of diamonds has led many to destruction."[7]

To the final sentiment few could have paid any attention; they were getting better advice from matchboxes. "Look prosperous," asserted William Fitz-Gerald, "that is the tacit order of the metropolis. You read this on the box of Diamond matches given away at the foot of the 'L' stairs. 'Wear diamonds' is another prompting on the label. To stint and save in New York is said to be the maddest extravagance of all, if a man is to win."[8]

Occasionally social excluders attempted to limit diamond display. The very rich Mrs. Paran Stevens piously opined that the poor should be "protected from having their eyes dazzled by the sparkle of diamonds in the ears of those who are more fortunate than they in the goods of this world." The social arbiter Mrs. Burton Harrison wrote of being "struck by the number of jewels worn in this community" and wondered disingenuously "why the wife of the lawyer, the doctor or the merchant in process of money-getting should assume a diamond tiara which seems by right to be the privilege of only the possessor of already accumulated wealth, who would be thought entitled to sit down with folded hands upon the high seats of American aristocracy." Some exclusionary moves were very complicated: the *Tribune*'s society reporter noted unhappily that some rich women were augmenting their personal displays by interspersing "Palais Royal brilliants" with real gems. "The general public is deceived!" trumpeted the *Tribune* before qualifying its disapproval: "If a woman has a certain number of fine jewels she may wear as many false ones in addition as she can without making the fact conspicuous. It is only in those who have no real ones that the display of false gems is vulgar."[9]

Meanwhile newspapers dazzled their readers with descriptions of persons occupying the "high seats" in society: Mrs. Ogden Mills wore "diamonds galore" and put on "prodigal displays of jewels," and Mrs. Astor appeared at a musicale done up like a human jeweler's tray, "dressed in black velvet and fairly covered with jewels. Strings of pearls covered her neck while a broad band of black velvet, on which were diamond ornaments, was around her throat. On the front of her corsage she wore a most brilliant ornament in the shape of a bowknot of diamonds, from which hung tassels of the same precious stones. In her hair she wore a black velvet bow with a diamond ornament." At the Bradley Martin ball she would appear in a "necklace and collarette of solitaire diamonds, and with the front of the corsage nearly hidden with a stomacher and ropes of diamonds." M. E. W. Sherwood, indefatigable apologist for society, wrote of Mrs. Astor in 1897 that "she was the first of our rich women to wear many diamonds, and she always looked as if they wearied her. Her heart was not in this world, or its pomps and vanities . . . philanthropy was her passion." It was, however, not Mrs. Astor's invisible heart but her diamonds that made the noise in public.[10]

In marked social contrast, a man on a low seat like the railroad-parts salesman Jim Brady turned himself into a notorious figure of fun and bad taste by wearing too many of his twenty thousand diamonds. On the fly, Wall Street messenger boys scanned the brokers' orders they were conveying, rushed to a bucket shop to buy or sell, and used their profits to deck their fingers with two-carat diamond rings. Other men settled for less public relationships with diamonds: the architect Stanford White, who was apparently uncomfortable wearing diamonds, nonetheless gave diamonds to women, and let on that he had his own valuable collection of jewels. The Rev. Henry Ward Beecher, whose public position suggested no appropriate connection with diamonds, preserved an outward piety in dress but carried a handful of precious stones in his pocket and fondled them when he was alone.[11]

Shoplifters—a category of consumer that crossed classes from top to bottom—made diamonds a special focus of activity in the 1890s. On January 14, 1894, Elizabeth Ryan, a Long Island City domestic worker, was running errands for her employer when she saw in a dry-goods store at Third Avenue and Fifty-ninth Street a diamond ring that she wanted badly. She literally consumed the ring by popping it into her mouth. The silk string holding the price tag—$1.97, or about $40 today—looped itself around a front tooth, leaving the tag itself dangling from Elizabeth Ryan's mouth. The saleswoman and the store detective watched her make "frantic efforts to draw it into her mouth or reach it with her tongue"; then they had her arrested. She was speaking accurately when she said, "I must have put it in my mouth in a moment of abstraction": a diamond had pure abstract consumer value unrelated to its appearance or quality. Nonetheless, if the diamond's show value held steady across classes, punishment for stealing the jewel varied considerably: when a Dr. Stickney, trying on diamond rings in a jewelry store, left with one on his finger and laid it to "absentmindedness," he, unlike Elizabeth Ryan, was excused as "a victim of a nervous disease"—and a cocaine addict.[12]

Across America, people understood the value of diamonds. Russell Conwell delivered over five thousand times a lecture titled "Acres of Diamonds" that sprang from a diamond-mine metaphor into an opportunity for Conwell to tell his audiences that "you ought to get rich, and it is your duty to get rich . . . you ought to have money . . . it is your Christian and godly duty to do so. . . . While we should sympathize with God's poor—that is, those who cannot help themselves—let us remember that there is not a poor person in the United States who was not made poor by his own shortcomings, or by the shortcomings of someone else. It is all wrong to be poor, anyhow." Higher-toned Americans had the likes of George Santayana to support their lust for diamonds: "Great as is the sensuous beauty of gems," wrote Santayana in 1896, "their rarity and price adds an expression of distinction to them, which they would never have if they were cheap. The standard of cost, the most vulgar of all standards, is such only when it remains empty and abstract."[13]

Diamonds functioned not only as personal décor, increasing the human capital of the wearer, but also as technology. Americans were hung with diamonds for exactly the same reason that gasoliers were hung with crystal prisms: both increased the diffusion of light. In a gaslit room, wall fixtures had to be positioned low—between five and a half and six feet from the floor—and a minimum of two feet from the ceiling. Unless perfectly managed, the gas flame tended to flicker; furthermore, gas jets ate oxygen and produced an odor, a misty vapor, and, depending on whether a person liked or hated the sound, a hum or a hiss. Gaslight had features peculiar to itself as a technology: gas fixtures pointed upward and thus threw their most effective light at the ceiling, and because gaslight tended to pool, it left parts of any room in considerable shadow. The color red looked especially good under gaslight, while dark blue, dark green, and all purple shades turned to gray. Gaslight's own color effects were usually described as *mellow* or *golden* and its intensity as a *dancing shimmer*, or a *mild radiance*, even as *sparkling* or *brilliant*. The latter, however, is an entirely relative term: a contemporary ten-watt electric candle-socket bulb comes closest to replicating the light produced by a fishtail gas burner.

Without doubt, the popularity of diamonds had a powerful connection to the nature of interior lighting in America between 1880 and 1900. The diamond's light-bending ability allows it to throw back almost all light that enters it. Like the prisms, reflectors, frosted glass, cut glass, and glass bead chains used to diffuse gaslight, the highly refractive diamond could not only diffuse the available light in a room but also light into visibility the face and person of its wearer. The more numerous and more faceted the diamonds, the more spectacularly visible was the person wearing them; an ornament of multifaceted diamonds was both a light resource and a source of visual pleasure in ways not replicated under the glare of late-twentieth-century lighting.

Across the final two decades of the nineteenth century, American cities gradually accomplished the changeover from gaslight to electric light. Because the sixty-watt electric lightbulb of 1893 produced a coppery glow that a contemporary observer experiences as about twenty watts, Americans in the 1890s massed lightbulbs and dotted festive spaces with dozens of small sparkling lights. Under 1893 electric lighting, diamonds looked pleasant enough but no longer shot back the light as they had under gas burners. If their technological function diminished, their value as personal décor stayed strong.[14]

In the spring of 1893, railroads failed, banks closed their doors, brokerage houses collapsed, well-known names declared bankruptcy, clerks and managers absconded with whatever they could pull out of failing firms, gold reserves streamed out of the country, money was tight, and the stock market stagnated. On May 5, 1893, two and a half weeks after Cornelia Martin's wedding, the word PANIC first appeared on the front pages of New York City newspapers, followed by WALL STREET'S BIGGEST CRASH on May 7. Across America throughout the summer of 1893, businesses failed and individuals were wiped out. Even the *New York Herald*, though devoted as ever to the interests of the society rich, whispered on July 30, 1893, that "the financial situation may seriously handicap the making of many good matches this summer . . . there is no chance for any profitable matrimonial engagements while this state of things continues, and the girls must resign themselves to a dull season. When the reaction comes, as it surely will, there will be no end of gayety, and the winter season of 1893–94 will utterly eclipse any former years."

The depression shut down diamond imports, but the value of diamonds held firm: jewelers announced that Christmas of 1893 was "just the time to buy diamonds, that it was better to buy diamonds than to give to the poor, and infinitely more businesslike to invest money in diamonds than to put it out at even those exorbitant rates of interest which scared quaking Wall street down into the depths of its varnished boots some weeks ago." For an extra 10 percent, Casperfeld & Company at No. 144 on the Bowery promised to take back all diamonds bought from it in case of financial disaster on the part of the purchaser; the *Herald* reporter thought the 10 percent "a trifling sum to pay for the temporary ownership of some of the blazing beauties that they offer." Across the next year, diamonds became a form of currency in rapid circulation: they were smuggled, sold, stolen, pawned, recut, and reset. Occasionally they had to be hidden: after the Lexow Committee investigation of the New York City Police Department went into session in 1894, "all the police department of the city of New York threw off their articles of jewelry" and denied that they had ever used to wear "large diamond rings."

In late 1894, twenty months into the depression, buyers of jewels were scarce, and busted European aristocrats trying to sell the family jewels in America had little luck. In December of that year, Prince del Drago, grandson of Isabella, ex–queen of Spain, arrived in New York bent on turning "his sole inheritance from his mother, a superb crown of diamonds and pearls, into bread winning cash, and hearing that there were now more tiaras and coronets worn in America than in all the courts of Europe, he brought his precious heirloom here and has put it up for sale. A rich woman coveted the crown dreadfully, but would only go half the price asked. Prince del Drago, in whom necessity had chilled the artistic sense, consented to sell half of it, and is still looking for a purchaser for the other half."[15]

When the Bradley Martins returned to America in November of

1894, the *Herald* hoped that their very presence, in a season when "no one seems to have life enough to jolly things up a bit," might give "a new impetus to the art of entertaining and create a reasonable hope that they would renew some of their former triumphs in that direction."[16] Since the crash, however, the rich had become more careful about flaunting their expenditures. Social arbiters, a decade earlier, had successfully urged the well-to-do to see the poor as responsible for their own ghastly situation; in the face of the terrors of the mid-1890s, they went silent. The once-dominant notion that the rich, by entertaining themselves, gave helpful employment to the poor had taken a beating, though it still had some life in it. In March 1893, in the newly opened Waldorf, the rich had a "fine supper" and a "brilliant and fashionable concert" in aid of sick children, none of whom were present. Six months later, however, ideas about charity were undergoing rapid transformation: in September of 1893, when a social projector proposing a charity ball for the poor of New York asked hotelkeepers to contribute toward a deluxe free supper for the rich attendees at the ball, he was refused. Hotelkeepers advised him that the problem of starvation would be more directly attacked if the poor themselves were to be invited to eat the deluxe supper. The idea that relief for the suffering poor was to be filtered through the intestines of the rich had lost some, but not all, of its social credibility. The Bachelors, in December of 1893, decided against throwing their annual ball and in favor of forwarding their ball funds to "the poor and needy." Five days later, however, after meeting at the Knickerbocker Club to discuss their social obligations, they decided their social debt to the hostesses of New York exceeded their responsibility to the needy, and so held their ball, "and a very smart one, too," after all. In a time when personal pleasure parted company with social benevolence, the Bachelors chose personal pleasure.

Further coming under attack were standard notions about relieving human suffering by distributing pale versions of major consumer icons. To this way of thinking, the poor who could not have banks of cut flowers might at least find solace in a single geranium. The Children's Potted Plant Society, a favorite charity of Jay Gould's daughter Helen, transported cartloads of potted plants to the Lower East Side and expected to distribute them in an orderly way, one plant apiece, to children who would act grateful. Society members were frightened and distressed when the assembled crowd of "grimy urchins" rushed the staging area, seized four or five plants apiece, and ran off to market them on the streets.[17]

Even though some financially comfortable people continued to claim that poverty was no bar to happiness, and that persons should aim to be pure, noble, and useful rather than rich, hard times touched everyone in ways worse than they had been touched by any of the previous panics of the late nineteenth century. Persons that the newspapers labeled "cranks" pressed their faces against the windows of Delmonico's and shouted threats at the patrons inside; the rich hired guards to watch their Fifth Avenue palaces and took their daughters to Paris to be married so as not to risk a show in New York. Cornelius Vanderbilt assessed the situation and decided to open a chain of pawnshops throughout the city. Unable to avoid the customary publicity for the big parties that some still threw, the rich at first took to giving watchmen's rattles as party favors, instead of the usual silver trinkets; when the rattles drew public scorn, they stopped giving party favors altogether.

The formerly rich, ashamed and embarrassed, struggled unsuccessfully to conceal their reduced circumstances. Some announced they were leaving town for expensive resorts but were subsequently seen at cheap watering places. Others lowered the shades to make the house appear closed for the season and then were spotted seeking some fresh air in Central Park. Even if the women escaped the city heat, the men stayed on and attended to what was left of business: at Newport in August of 1894, according to the *Herald*, "the curtain is up, the lights are fully turned on, the pretty white and gold dancing hall at the Casino is ablaze with diamonds, but the only drawback is the paucity of dancing men. The musicians play, sets are formed for the old fashioned lanciers and quadrilles, and the women festoon a mark on their arms to designate to the female partners they have taken the place of male dancers, and so the dance goes on." The record-smashing heat waves and ferocious winters of 1893, 1894, and 1895 were "death-dealing scourges": thousands of New York City's poor starved, fell out of windows while seeking a breath of air, and froze to death. In the single week of July 17–23, 1893, 607 children under the age of twelve months died in the tenements of New York City. The *Herald* thanked those "New Yorkers departing for summer resorts who generously remembered those who can't escape the heat" by donating a few dollars to the *Herald*'s Free Ice Fund. The question of what money was for began to arise with frequency.[18]

Any hopes that the Bradley Martins could restore cheer to a scene of business stagnation and personal despair faded when the Martins' problematic elder son, Sherman, twenty-six years old, also arrived in New York and engaged a suite at the Morton House at Fourteenth Street and Union Square. Three years earlier, Sherman Martin had made news when he was "entrapped into a marriage with a woman of the town, in London." He had been extricated from that marriage but had developed a consumer problem. Despite having recently taken the cure at the Hartford Retreat for the Insane, in New York Sherman Martin went on a binge and "drank freely" in public for two weeks. At six in the evening of Friday, December 21, Sherman Martin reeled into the Morton House café "arm in arm with a boon companion" and ordered cocktails. He was refused service. For a man who, like his parents, was positioned in the economy as a major consumer, this treatment was intolerable. While raging at the servitor assigned to eighty-six him, Sherman Martin turned livid, gasped for breath, and fell out of his chair. He died four hours later. In the face of the sensational publicity that followed on his death, the Bradley Martins fled to Albany, their childhood home. There, on January 5, fire broke out in the hotel where they were staying. The *Herald,* paying its usual homage to the rich and their goods, reported that "Mr. Martin hurried his wife and son to a place of safety, leaving a crippled elevator boy to guard their personal effects. Rushing back to his rooms to save what he could, Mr. Martin nearly lost his way and his life in the blinding smoke, and might have perished but for the coolness of the boy. Of course, Mrs. Martin had no jewels with her, as reported, but she and Mr. Martin left valuable sables, dressing bags and clothing in her apartment, which were all happily saved." The Bradley Martins, always uneasy with publicity that they did not control, stayed away from New York City for two years.[19]

When the Bradley Martins returned to America in 1897, for what would be their final campaign as public superconsumers, they triggered an international discussion of money, class, style, and appropriate expenditure. Perhaps they were, like Henry James's expatriate Frederick Winterbourne in *Daisy Miller*, "booked to make a mistake" because they had "lived too long in foreign parts." In the winter of 1897, if the Bradley Martins were aware of unemployment, starvation, infant mortality rates, death from exposure, and widespread despair in their native land, they had certainly neither felt nor seen any of it. Never attuned to the lives of anyone but themselves, in 1897 the Bradley Martins made it apparent that their icons were not only still in place but in fact required even greater outlays of human energy and dollars for their satisfaction. They did, however, exhibit a little uncertainty about the nature of their next big show.

After considering a musicale, and pondering erection of a temporary corrugated-iron ballroom, Mrs. Bradley Martin suddenly decided that the Waldorf, with its massive entertainment resources, would be her venue. She chose February 10 as a party date, and sent out twelve hundred invitations to a costume ball. Her haste, though she did not say so, was tax-driven: by staying in the United States no longer than three months, the Bradley Martins hoped to claim a nonresident tax exemption on their New York City properties. Her guests would have to rush to accommodate the Bradley Martins' loathing of taxes. The family fortune was hers, as the newspapers carefully explained at the time:

> Bradley Martin was not wealthy when he married Albany native Isaac Sherman's only child, but when Isaac Sherman died he left $4,000,000 [equivalent to between $60,000,000 and $80,000,000 today] to his daughter. Now $4,000,000 is not a great principal to splurge on in New York society, but every penny of it was invested in big dividend-producing properties. There are persons in the swim of society whose millions count five times the fortune Isaac Sherman left who

could not encompass with their incomes one-half the expenditures of the Bradley Martins.[20]

The invitation itself read "Mrs. Bradley Martin requests the pleasure
of . . ." and announced "Costumes of the 16th, 17th and 18th centuries
de rigueur." Hearst's *Journal* was astounded: "Where does Mr. Bradley
Martin come in? Does he cut no ice on this occasion?" A woman who
not only controlled her own money but also directed others in spending
theirs was a phenomenon that bothered the *Journal*, a newspaper that
generally clung to the notion that woman spent and man paid the bills.
The next day Cholly Knickerbocker took up the subject.

> Up to the present time Mr. Bradley-Martin has cut no ice in
> these ball proceedings. His name does not appear on the invi-
> tations, and he is not in evidence anywhere else that I can dis-
> cover, but he will probably be around somewhere when the
> bills come in, and I presume that he will go to the ball. Of
> course, Mr. Bradley-Martin's gallantry would tempt him to
> go as Sir Walter Raleigh, who threw his cloak in the mud for
> Elizabeth to step upon, while his affection might suggest
> Dudley, Earl of Leicester, whom Elizabeth regarded as a
> good thing to have in the house, but I am of the opinion that
> he ought to go as Henry VIII. Of course, Mr. Martin resem-
> bles him in no particular whatever, except that Henry termi-
> nated the proceedings on the Field of the Cloth of Gold with
> a fancy dress ball that created as great a sensation in its day as
> the Bradley-Martin affair is doing now. One notable result of
> Henry's ball was to beggar many of the French and English
> nobles, who bankrupted themselves in order to make a great
> show thereat, an example that certain chappies of my
> acquaintance are most fearful that they will follow on the
> 10th proximo.[21]

Perhaps Mr. Bradley Martin had been erased in service of Mrs.
Bradley Martin's illusion of herself as an occasional visitor from
abroad who invariably outclassed the locals; Bradley Martin lost even
his first name when she encouraged the newspapers to hyphenate

Bradley-Martin. Because she had the money, she claimed authority, in 1897 New York, to register her demands. According to Richard Welling, one of the rare invited persons who left a narrative of the event, "not only was the censor to have authority to reject people coming in any old fancy dress, but the guests were requested in their acceptances to state what court dress they intended to wear." By demanding that "the court" be recovered at her party, Mrs. Bradley Martin suggested that in America a court—or her court, at least—could be created if she could just bully enough "guests" into giving it a go. So she threatened her guests with the possibility of public humiliation at the very door of the Waldorf: if guests were not costumed as she had demanded, an authority figure representing her wishes might turn them away in full view of other party guests and—worse yet—the expected jeering street crowds.[22]

Mrs. Bradley Martin had still further uses for her guests. Always attuned to the prospects for publicity, she described to newspaper reporters the decorations and the drill to be followed at the ball; then, taking an unprecedented step, she also handed to the reporters lists of attendees and their costume choices. On succeeding days she distributed detailed updates of those lists. Such publicity efforts ensured that the Bradley Martin ball would occupy at least two columns of newsprint per day. At first the publicity seemed to backfire just slightly, when some guests who loathed seeing their names and costume notions in print decided not to attend after all. Then, however, repercussions enlarged: ministers from the pulpit and college debaters and a range of persons equipped with social ideas began to publicly register criticism. On the one hand, the Bradley Martins did perhaps the only social service of their lives when they launched a national conversation about the meaning of wealth-holding and consumption; on the other hand, to some listeners the conversation itself was too dangerous to hold. Frightened people who knew they were helpless to stop the Bradley Martins tried instead to stop their critics by claiming that criticism further heated an already "fevered" social situation. Who, in this tangle, was more dangerous to society: the suffering poor, the social critics, or the insensitive rich? Should the Bradley Martins be left untouched while they exacerbated social tensions, or should their critics exacerbate social tensions by pointing out to the Bradley Martins and their guests just how insensitive and dangerous a group they were?

Among the first critics to weigh in was the Rev. Dr. William Stephen Rainsford, rector of St. George's Protestant Episcopal Church, appointed to his post by Senior Warden J. Pierpont Morgan, and a local celebrity in his own right. Rainsford, in the context of the 1890s, looked like a social activist: while ministering to a congregation of the very rich he had expressed strong views on opening up St. George's to the poor. Rainsford's expertise at publicity nearly rivaled that of Mrs. Bradley Martin. On January 21, Rainsford allowed himself to be "interviewed" on the subject of the upcoming costume ball, and he let out that on Sunday, January 24, he would be making a major statement from the pulpit. That Sunday, "hundreds stood in the aisles while cushions were brought from the vestry room and placed on the stone steps of the chancel, to accommodate a part of the overflow." Rainsford was no socialist; he "urged his hearers to avoid all outward display of riches that might breed discontent among the poor," and after reminding them that "the entire miseries of the world, in these times of rapid communication, are before us every day," he offered, as a solution, "being kind." Several days later he told the *Tribune* that he had "advised some of his parishioners against countenancing the Bradley Martin ball by attending it." The *New York Times*'s society reporter scoffed at Rainsford's "grand-stand play," called it "ill-timed and discourteous," and sneered that "he has not had any particular amount of notoriety during the last year." How much effect Rainsford could have and what direction that effect might take were both arguable. In 1896, in a famous essay titled "The Expenditure of Rich Men," E. L. Godkin mused that "one of the odd things about wealth is the small impression the preachers and moralists have ever made about it." In the 1890s, Godkin thought, "the rich man is now allowed to have wealth, but the ethical writers and the clergy supervise his expenditure closely."[23]

For many Americans, however, a man who "ministered" to the rich had high credibility as a social critic—second only to that of a really rich man. Furthermore, several Manhattan clergymen were in the advantageous position of regularly having their Sunday sermons reported on in Monday's newspapers, and on the following Sunday, others among them weighed in on the subject of display. At Temple Emanu-El, Dr. Gottheil asserted that "whatever wealth buys or creates for the embellishment of life is intended to be seen and admired. What other use is there for wealth?" He suggested addressing the question of "how any wealth was got" rather than how it was spent. Dr. Madison C. Peters at Bloomingdale Reformed Church, after admitting that "I wish I was a millionaire myself," nonetheless labeled all display "tawdry" and was especially critical of rich families who "parade themselves before the public, ape the fashions of Europe, and marry their daughters to titled mendicants." At the Church of the Divine Paternity, the Rev. Charles Eaton worried over "problems of increasing wealth" and "unrest in the body politic that social agitators may use to influence the ignorant to anarchism." Speaking to the Standard Oil Company congregation that attended the Fifth Avenue Baptist Church, the Rev. Dr. W. H. P. Faunce tossed both diamonds and flowers into a category he labeled "non-productive expenditure," or, more directly, "waste, crime, and sin." The sociologist John Graham Brooks told his audience at the Berkeley Lyceum that "the idea that the rich benefit the poor by the wildest possible expenditure of money is a theory that has been cast aside by all sane economists as the most disreputable of all fallacies," but the economist George Gunton told his audience at the School of Social Economics that the Bradley Martins' expenditures would put so much money into the pockets of even the "humblest laborers" that he himself hoped for "the chance to one day get into a ball" and do social service thereby.

Inside of a week, the Bradley Martin ball became, according to the newspapers, a "universal and engrossing subject of interest and discussion" across both America and western Europe, and Mrs. Bradley Martin herself became "a National Issue." Newspaper editors in London and Paris "made special preparations for cable reports." Disgusted newspaper readers, and some whose reactions went way beyond disgust, fired off letters to the Bradley Martins, the known attendees, and the management of the Waldorf. Letter-writers threatened to blow Mrs. Bradley Martin's house to bits on the afternoon of the ball, to explode a bomb in front of her as she stepped from her carriage at the Waldorf on the night of the ball, and to blow up the ballroom itself. As a result, all windows on the two lower floors of the Waldorf were boarded up, all entrances were draped to deny the uninvited even a glimpse of the interior, and orders went out that no locals would be allowed to "stand and look at the hotel or stand and talk." Elaborate plans were laid with the police department to cordon off side streets in the area of the hotel, and there was even talk of positioning sharpshooters atop the hotel roof. Mrs. Bradley Martin gave "the most rigid orders" that there should be "no chance whatever for outsiders to get a peep at the pageant"; she hired a hundred detectives to line the corridors of the Waldorf posing as "lackeys in gorgeous liveries." How the detectives felt about costuming themselves as what Erving Goffman calls "craft-bound servitors" is not on record, but the practice of dotting private social events with detectives always astonished foreign visitors. "What a barocco idea!" wrote Guy de Soissons in 1896. "To have detectives at the balls and at the weddings!"[24]

By first publicizing the costumes, floral displays, and décor, and then publicizing their restrictions on who could see the show, the Bradley Martins made a sizable mistake. They unpacked their publicity-hungry selves and their monstrous display into a scene of human misery, they told everyone how much there would be to look at, and then they forbade their audience a look. In a culture hungry to consume visually, a culture where the poor took their children through toy stores simply to see toys that only the children of the rich would ever play with, the Bradley Martins planned to cheat the audience of the promised show. Threats frightened them, and fear made them exclusive. Furthermore, they failed to comprehend a shift in social thought that had occurred since 1893. The social idea that the rich could spend their money as they wished still dominated, but it had acquired a corollary: big spenders had to allow others to see the display their money had bought. Theater celebrities, who were about to supplant the social rich in the public eye, understood this corollary perfectly. Lillian Russell knew, after Jim Brady gave her a gold-plated bicycle, that having got it she should let others see it. So, on Sunday afternoons, the American Beauty majestically pedaled her golden bicycle through Central Park. The Bradley Martins, however, had "determined to exclude the public . . . even the guests of the hotel . . . from so much as getting a glimpse of the splendors." Far from engaging in "conspicuous consumption," the Bradley Martins were engaged in publicizing concealed consumption, in making Americans hungry to see but then dropping a curtain before the desired vision.[25]

Mrs. Bradley Martin fished for twelve hundred guests and not quite seven hundred rose to the bait. Over the course of the next few weeks, they pored over illustrated books, lingered in art galleries, rushed costumers and perruquiers, ransacked pawnshops for extra jewelry, and pleaded with theaters and the Metropolitan Opera for the loan of "court costumes." Some persons crossed class levels in their effort to costume themselves and were surprised to "learn that the lower east side of the city was dotted with costumers who catered to the masquerade and fancy dress loving element among the residents in that section." The taste for costume that ran across class levels had nothing to do with any influence trickling down from the rich, for "strange to relate, the first to find these costumers discovered that they had not heard of the Martin ball, and so obtained some very good costumes at a figure which seemed startlingly low. After a few days the down-town costumers, surprised by this sudden rush of new custom, discovered the cause and up soared their prices, also, but only relatively to those of the up-town outfitters." Some potential guests dropped out when they tried on their costumes and decided they looked not only silly and awkward but "inauthentic." Male guests accustomed to a life of just two costumes—either the black business suit or standard evening dress—borrowed women's silk hose, worried about their legs, and agonized about shaving off their mustaches in the service of "authenticity." In truth, however, the Bradley Martins had brought America to a historical confusion point not over authenticity but over matters of money, class, and consumption. They had described their game in detail to millions of newspaper readers and they had asked twelve hundred people to pay the price of playing it; five hundred had refused to play, but seven hundred, many of them uncertain about available grounds of resistance, had been beaten into submission.

Many ball guests booked rooms at the Waldorf in order to protect

themselves and their goods from the crowds expected to gather outside the hotel. Others, on the evening of the ball, fought their way
through throngs of onlookers who had settled in for five blocks
around the Waldorf. The uninvited attempted to peer into stalled carriages just as they had four years earlier at Cornelia Martin's wedding,
and impatient guests who left their carriages to walk the last block to
the hotel were met with jeers and catcalls and "Say, mister, you forgot
your pants" and "Off with his head" from the assembled locals. Richard
Welling wanted to see the Bradley Martins' show but evade their
European-royalty game, so he dressed as Miantomah, chief of the Nar-
ragansetts, and found himself to be a special focus for ridicule when
the height of his headdress forced him to take an open victoria instead
of a closed cab. "Imagine," wrote Welling, "the grilling, the yelling,
and the hooting the hoodlums gave me!"[26]

Overall, however, the dense crowd waiting outside the Waldorf saw
very little, and "there were abundant exclamations of disappointment
from the moving throng." Furthermore, the guests themselves were
neither as numerous nor as interested in the proceedings as the
Bradley Martins might have hoped. According to the *Times*,

> Little more than half of the twelve hundred invited were in
> attendance. Of those that came, also, quite a number left early.
> Many seemed to have put in an appearance simply out of
> curiosity, to witness the really superb decorations of the ball-
> room. Within half an hour after the beginning of the ball, guests
> began to leave the place, some for their homes and others to
> wander about the hotel corridors. That the visitors had a set
> purpose in leaving early was manifest from the fact that fully
> half the carriages were ordered for before the time set for
> the cotillon.

The late hours that the society rich had been keeping for the previous two decades were coming to seem less tolerable; moreover, the
newspapers had helped them to feel old by listing the goodly throng
claimed by Old Mortality since the last Bradley Martin do in 1893. To
struggle into a difficult costume and a heavy wig, attend a pre-ball dinner, arrive at the Waldorf at 11 P.M., eat a heavy supper at 1 A.M., and

still be standing when the cotillon began at 3 A.M. constituted a pleasure schedule that made serious demands on aging bodies. At least as early as 1893, projectors of various balls, noting that "for the man of affairs these late balls are out of the question," had begun to specify that their *affaire* would end "exactly at midnight." Furthermore, for three weeks—in an America where three weeks was a long time for the maintenance of public attention on any event—the Bradley Martins' publicizing of every aspect of the event had been so intense and the battle over the appropriateness of the ball so loud and confused that interest may have been on the wane well before the ball itself occurred. Perhaps it was quite enough to see Mrs. Bradley Martin, her diamonds, the flowers, the lights, and the velvet costumes. That was what she offered—a sight. Her mistake lay in having been too exclusionary about who got to see the sight. Suddenly, in February of 1897, it was not only a mistake for wealth-holders to put on a display but it was also a mistake not to put on a display.[27]

Within the Waldorf on February 10, the cluster of desirable goods was complete: hostess and guests were dressed in velvet and loaded with jewels, there were red velvet balustrades, crimson satin curtains, clusters of incandescent lights, and, depending on which newspaper one read, either seventeen hundred or six thousand orchids, and five thousand to thirteen thousand roses. The flowers, in combination with the increasing warmth in the rooms, produced a narcotic effect. The late nineteenth century was, as many remarked at the time, a Floral Age; hothouses in the vicinity of New York City constituted a $180 million business, and there were innumerable florists' shops and street vendors of flowers. Not even this local bounty, however, could sate Mrs. Bradley Martin's appetite for flowers and greenery. According to the *Times,* "a great army of poor folks in Alabama have been engaged in gathering clematis vines for the affair," and florist J. H. Small told the *Herald* that the other greens "consisted chiefly of smilax, galax and 'royal' asparagus vines. The smilax came from the swamps of Florida. Our agent down there at once engaged fifteen men, their wives and children, who plunged into the swamps and got this. It took four days and the continuous use of six wagons."

The Floral Age had nothing to do with gardening. The favored flowers and greenery were either hothouse goods, sold by the piece, or they were very difficult to obtain, requiring whole families of day laborers to retrieve them from distant southern swamps. British visitors to America, accustomed to the gardening passion in their native land, marveled at the commodity nature of flowers in New York City. In 1887, Catherine Bates remarked that "flowers form a terrible item in the expense of social life to those who are not rich enough to hold dollars in contempt." Thirteen years later, Philip Burne-Jones observed the phenomenon still going at full strength: "Flowers," he wrote, "chiefly raised in hot-houses, have an enormous sale in New York, and fetch preposterous prices. Yet with all their love of flowers, people seldom have them growing in their windows, and the dreary streets are enlivened by few bright boxes of plants or flowers." Across social levels, city people desired the same floral look that Mrs. Bradley Martin strove for. In his 1890 *Manners, Culture and Dress of the Best American Society*, the kind of advice book that a struggling middle-class person might read, Richard Wells devoted his entire "Home Decoration" chapter to ferns and flowers, beautiful things that, Wells complained, were too often, "like everything else in this lower world, regarded as the sole perquisite of the rich."

Flowers, like birds and small mammals, were in 1890s New York not nature but decorative objects. In consumption terms, moreover, flowers were a "pure" purchase, "for the expression of pure, dignified sentiment," according to the florist Charles Thorley; he asserted that "into no other business does so little of the immoral ever creep." The claimed "pure pleasure" of flowers had, however, got complicated by the familiar and always unstoppable American propensity to enlarge objects. In 1893, young women appeared carrying bouquets of two thousand violets, and chrysanthemums ten inches in diameter were the fashion in 1894. In 1897, the social value of orchids was in the ascendant, and the *Herald* advised any man who wished to be thought "well dressed" to wear "orchids if possible" massed into a boutonnière the size of "a perfect young cauliflower."[28]

It was among mauve orchids—"floral aristocrats chosen to grace this occasion because of their acknowledged standing in the world of flowers"—and "fifty handsome boys dressed as pages" that Mrs.

Bradley Martin positioned herself on the evening of February 10. Mr. Bradley Martin seemed to have receded further into the background while Mrs. Bradley Martin seized the entire foreground: she appeared as Mary Stuart in black velvet lined with cerise satin, lace that she valued at today's equivalent of thirty thousand dollars a yard, and "many of her splendid jewels": a stomacher, necklace, and head ornament, all of diamonds. She had previously worn this dress in 1883, at the fancy-dress ball given by Mr. and Mrs. W. K. Vanderbilt for the Duchess of Manchester. This time Mrs. Bradley Martin played a dual role: she was both wealthy giver and royal guest. She stood on a slightly raised velvet-covered dais to receive her guests; Mr. Bradley Martin stood off to her right and below the dais. She was not averse to some homage: she and John Jacob Astor led a quadrille as queen and king, and then seated themselves while the other dancers "backed slowly to chairs at the sides of the room," bowing to her as they went. She was a dynamo of consumption whose talent for expenditure guests were forced to recognize. The potency of her show enwrapped even the most reluctant of her guests. A male guest who was reluctantly done up as Louis XV found happiness when he shed his Marie Antoinette and located the supper room, wherein forty waiters offered unlimited supplies of terrapin, canvasback duck, British pheasant, and liquor. With diamonds and flowers and velvets and lights Mrs. Bradley Martin couldn't hook him, but with luxury comestibles she got her man. He was so happy that when a "beastly old friend of the family" tapped his shoulder and whispered that Marie Antoinette thought it time to go home, Louis XV was appalled: "What? Home? To go back to tweeds and long trousers, after all this satin and silk? Home? To the end of the nineteenth century and bills and all the rest of the old grind? Never!"[29]

For days after the ball, intense public effort was bent on understanding and explaining the event. The newspapers battled to set a dollar value on the whole affair, then as now a common American stab at meaning; the *Tribune*, in a series of defensive guesses, managed to whittle the estimated price of the ball all the way down to a hundred thousand dollars, but the *Herald* stuck with its estimate of three million dollars changing hands. The expenditure angle was a difficult point of attack in an atmosphere wherein everyone agreed on the beauty and value and desirability of jewels, flowers, velvet costumes, and twinkling lights. If most everyone was willing to spend on the same goods, then neither the expenditure nor the goods offended; offense lay in the idea that expenditure was appropriate to but one social class—to the Versailles class specifically. The class paraphernalia with which the Bradley Martins had surrounded their ball, combined with their determination to send an exclusionary message, constituted the problem at hand. George Boldt, manager at the Waldorf, insisted that exclusionary security precautions had been necessary because, he estimated, each woman guest had been wearing a hundred thousand dollars' worth of jewels. "The public," said Boldt, "can readily understand what would have been the result of a panic among an assembly of this sort and why it was so necessary to guard against anything which might approach an unexpected sensation." Boldt's explanation failed, and not only because it was offensive to the general public. Most of the jewels worn to the ball were paste and everyone knew it. In public crushes, as one partygoer put it, "Diamonds are wearing on the mind."

Many clung to the notion that the ball had furnished desperately needed employment. The florist J. H. Small, the costumers, and the perruquiers stepped up to repeat the idea, now under so much attack, that expenditures on costume "were of much benefit to the working-men and working women who were employed," and that "had it not been for Mrs. Bradley Martin's ball, these persons would unfortunately have had no work." It was obvious, however, that only a few had benefited, and at that for a few weeks only. Furthermore, some ballgoers thought they had been taken advantage of by hairdressers and costumers with "too great a propensity to cut a whole lot of hay while the Bradley Martin sun shines." The *New York Times* society reporter announced that critics of the ball were "sentimentalists" whose remarks could very well have stirred cranks into attempting violence on the Bradley Martins. The *Herald* asserted that the affair had not even been about money: "Such a result was not reached by the mere spending of money, but by the exercise of *exquisite taste*. The idea of limiting the costumes to the most lavish periods of history was a

happy inspiration, and in spite of much cavilling and ill-judged cen-
sure, the ball has proved as successful from a utilitarian as from a social
point of view."

The *Times* editorial page thought it just as well that the money had
been squandered in New York as squandered elsewhere, but was critical of such "rapid consumption within a small circle and for a short
period." And although the *Times* was certain that possessors of wealth
had a "right to expend it in their own manner," it was also certain that
"there are much better methods of distributing superfluous income
than to burn it in a round of fashionable gayeties that give employment
to milliners, florists, and flunkies the year through." Perhaps the weak-
est point in the argument that the Bradley Martin ball had furnished
employment was its distinct class separation: it defined the many as
producers of goods and services that only the few would consume. In
America, at the close of the nineteenth century, the many rejected a
dreary world wherein they merely stood by and awaited a chance to
enable the consumption-crazy few. The many wanted both to consume and to get a good look at what others were consuming.

The *Times* used another voice to say more when it reprinted a critique from the London *Daily News:*

> Some ask why Mr. Bradley Martin should not be, so to
> speak, sold and given to the poor; others, with Lord Penrhyn,
> why a man should not do what he likes with his own. The dis-
> cussion is flavored with fearful reports of eighty thousand
> persons kept alive on rations in Chicago. With some, Mr.
> Bradley Martin is a philanthropist, not to say a patriot,
> whose luxury furnishes employment to the poor. With others,
> he is little short of a madman not to wait for better times.
> America is a very old society, reckoning its age by its ideas,
> and there is nothing more characteristic of societies of that
> kind than the defiant animation with which they dance when
> anybody ventures to whisper that they are on the edge of a
> volcano. Mr. Bradley Martin and his guests have no belief in
> the volcano. The discussion over the Bradley Martin ball
> shows that the mind, if not the conscience, of wealthy and
> pleasure-loving America is ill at ease.

The Bradley Martins themselves tried for wiggle room by claiming that they had spent far less on the ball than their guests had, but that attempt to shift consumer blame did not mix well with their previous assertion that they had a right to spend as they wished; furthermore, it contradicted utterly the claim that their expenditures were massive enough to "percolate down" to the benefit of everyone. Then they let on that the floral decorations had been "distributed to policemen to take to their wives, and those left today will be distributed among some of the hospitals," but the trickle-down of wilting leftovers from the pleasures of the rich was becoming, in 1897 America, somewhat distasteful, a misunderstanding of the cultural give-and-take that was required and desired.

If discussion of the Bradley Martins' doings is reframed within Claude Lévi-Strauss's definition of culture as "imaginary resolutions to real contradictions," then the real contradiction in 1897 lay in how there could be such poverty and human misery in a land of plenty where so many held so much wealth; to this contradiction the Bradley Martins' party invocation of European court life was a perfectly unsuitable "imaginary resolution" in that it mirrored the social situation rather than reshaping it. A ball guest who remained anonymous mocked the idea that Mrs. Bradley Martin had been doing her bit for society: he told the *Herald* that he hadn't minded the street crowd because he had "consoled himself with the thought that in attracting such attention I was disseminating money among the poor."

Furthermore, uninvited locals refused to accept the notion that their lives should be disrupted by Mrs. Bradley Martin's party plans. Mr. S. E. Franklin of 261 East Forty-first Street went to the *Herald* office to complain that he and his sister had not been allowed to pass through police lines. "They were very dictatorial," he said of the police, "and ordered me and my sister back, refusing to let us pass along a public thoroughfare. I am not a socialist or anything of the sort, but it is the enforcement of just such arbitrary rules for the sake of a favored class that causes a lot of this demagogical agitation. I do not see by what possible law a citizen can be kept from walking along one of the public streets." The Central Office of the New York City Police Department, having been bought off with Mrs. Bradley Martin's left-over flowers, countered S. E. Franklin, and played to American racist attitudes, by claiming the police had kept such perfect order that they had been obliged to arrest just one person, George Walker, age thirty, "a colored man who was pushing his way through the crowd in front of the Hotel Waldorf and shoving against pedestrians in a suspicious manner." That he could have been the only New York male to so behave was beyond belief.[30]

Throughout the twentieth century right up to the present moment, a somewhat conflicted story has passed, unchecked, from book to book: supposedly criticism of the ball forced the Bradley Martins to flee America in shame, never to return, and although their insensitive and filthy-rich friends gave them a huge farewell banquet at the Waldorf, they were further punished by an increased tax assessment. The story suggests an unusual American effort to punish expenditure. The Bradley Martins did not flee America in 1897; they had been nonresidents since 1882, and they continued to be so. When they returned in 1899 they themselves, not their friends, threw a dinner on May 16 at the Waldorf-Astoria for eighty-six of the social rich. The Bradley Martins spent and behaved as always: they smothered the Astor Gallery and its ten-by-fifty-foot dining table in sweet peas, white lilacs, dogwood, and American Beauty roses. The orchestra, according to the *Times*, played "a popular negro song, 'If You Haint Got No Money You Needn't Come 'Round,' which seemed particularly to please the fancy of the guests." The rest of their time in America was devoted to Mr. Bradley Martin playing the role he had played six years earlier with his daughter's wedding dress: avoiding assessment, in this case property taxes of fifteen thousand dollars on 20 and 22 West Twentieth Street, a double-wide that had been valued at two hundred thousand dollars. Bradley Martin took an oath that he was an American citizen, he took another oath that he was a British subject, he called the assessment "exorbitant," and he produced as proof of tax exemption some of the newspaper clippings that he had always loved and believed in. He fought his tax assessment all the way to the New York State Supreme Court, where he lost. Only then did the Bradley Martins sink out of sight, sending up just one bubble when, in 1901, Mrs. Bradley Martin, as lost as ever in her royal fantasy, remarked again on what money ought to buy when she spent today's equivalent of twelve million dollars for a diamond tiara, a replica of Empress Josephine's diadem, and announced that she would be displaying it at the coronation of Edward VII.[31]

For better than two decades, Americans across the country had followed the doings of the society rich as if their affairs were worthy of attention.[32] That era of social celebrity for the rich had, in 1897, not much more than another decade to run. Already Americans had begun to make celebrities of those who entertained the general public rather than of those who entertained only each other. Theater and musical-stage performers, artists, and architects were proving that they could manipulate the icons of the time to even better effect than the likes of the Bradley Martins, and—most important—could do so without condescending to or disgusting the public. If the society rich were on the fade as celebrities, the specific grouping of goods that they knew how to produce was still glittering, and others would be far more expert at manipulating it.

The *Herald* offered Mrs. Bradley Martin the cover of exquisite taste for her doings, but she was less a taste leader than a consumption goddess. The *Times*, on the other hand, offered her the cover of *art* and the *artistic* and even claimed that, because she created nothing, she had offered her fellow Americans "a lesson in the art of living." It was too late for Mrs. Bradley Martin to conceal her greed and her bullying self behind *art*, but in the twilight of the American nineteenth century there was in fact no better cover for desire and expenditure than *the artistic*. Art and décor had shown themselves capable of furnishing and explaining the entire content of a life. Possibly one could have and shine, do anything and be anyone, so long as the effect could be called *artistic*.

NOTES

1. All Bradley Martin wedding and ball materials not individually cited are drawn from the *New York Times,* the *New York Herald,* the *New York Sun,* and the *New York Journal,* 1–20 April 1893 and 23 January–12 February 1897.

2. The Earl of Craven's brilliantly colored tattoos were spotted during his

daily swims at the Racquet Club. On 18 April 1893, the *New York Times* gave the tattoos front-page above-the-fold coverage, and explained that "tattooing is nearly general among swaggermen in England, and it is declared that there is any number of men in this city who have been beautifully decorated and who are simply walking picture galleries. There are no tattooers in New York worthy of the name, and only those young men who have had the good fortune to have made the Eastern trip are able to exhibit evidences of this new fad. Reports from Paris state that it is also becoming fashionable with women, and that the Princess de Leon, the Princess de Sagan, and other well-known leaders are being gorgeously tattooed by skilled East Indian artists."

3. According to Burrows and Wallace, Bishop Potter was "pastor to the smart set . . . he loved officiating at weddings between English aristocrats and the New York rich, and when he traveled to church conventions he rode in J. P. Morgan's private railroad car" (1087).

4. *New York Herald,* 1 December 1895, 13 December 1896, and 20 December 1896. No one familiar with transatlantic marriages could have thought that the Martin/Craven union was risk-free. Under the headline "Noble Now, but Still Unhappy," the *Herald* noted on 2 December 1894, "The great argument generally advanced against marriages of American women with foreign noblemen—of course, all noblemen must be foreign to this country—is that within a few weeks or years, as the case may be, the couple disagree, the bride returns to her parents and the groom negotiates drafts upon his father-in-law or otherwise attacks the family exchequer."

5. Between 1880 and 1910, 817 American women married British and European titles. The phenomenon is thoroughly discussed in Ruth Brandon, *The Dollar Princesses*, and Maureen Montgomery, *Gilded Prostitution*.

6. In *The World of Goods*, the anthropologist Mary Douglas encourages an end to Veblenesque disapproval of a high value set on the acquisition of material things. Douglas suggests examining both the things and the nature of acquisition itself. "All goods," writes Douglas, "carry meaning, but none by itself. One physical object has not meaning by itself, and the question of why it is valued has no meaning either. The meaning is in the relations between all the goods" (40).

7. *New York Herald,* 8 December 1895; Richard Wells, 434. See Edith Wharton, *The Age of Innocence,* either the novel itself or the 1995 film version, for a rendering of the social rich enjoying the "Jewel Song." "The frequency with which Gounod's *Faust* is repeated at the Metropolitan Opera House suggests the question whether it is destined to immortality," wrote the *Herald*'s arts writer, "but it is pretty generally conceded that all the immortal music ultimately dies of old age" (24 December 1896).

8. Fitz-Gerald, 8.

9. *New York Herald,* 29 January 1893, 24 September 1893. In the twentieth century, the price of diamonds was driven up to artificial levels by hoarding on the part of the diamond monopoly.

10. *New York Herald,* 21 January 1895, 30 December 1894; Sherwood, 185.

11. In *The Life of Henry Ward Beecher,* Joseph Howard discusses Beecher's collections of books, etchings, engravings, and paintings, and then goes on to address his other acquisitive habits: "It has been said that Mr. Beecher was fond of precious stones, and a collector of them. To a certain extent that was so, but not sufficiently so to account for any noticeable expenditure. He delighted in good horses, and drove the best he could procure. Although extremely modest in his attire, his clothing was always of the finest material. He was particularly nice and tenacious about the quality of his clothes. Many an hour did he spend with his old friend Knox, discussing silk, felt, beaver, human nature and theology" (637–638). Between 1847 and 1887, Beecher made at least a million and a half nineteenth-century dollars.

12. *New York Herald,* 28 October 1894, 14 January 1894, 18 February 1893, 24 December 1893, 25 December 1893.

13. Conwell, 20–21; Santayana, 130.

14. Sheldon, 137, 114–115; Penzel, 103–107; Gerhard, 108–141; Keating, 36.

15. Lexow Committee, V:5545; *New York Herald,* 19 November 1893, 6 January 1895.

16. *New York Herald,* 9 December 1894, 16 December 1894.

17. *New York Herald,* 24 December 1893, 29 December 1893, 31 December 1893.

18. *New York Herald,* 24 September 1893, 11 June 1893, 23 February 1894, 24 July 1893.

19. *New York Herald,* 22 December 1894, 6 January 1895, 23 July 1893.

20. *New York Journal,* 23 January 1897.

21. *New York Journal,* 24 January 1897.

22. James, *Daisy Miller,* 22; Welling, 270. The *New York Sun,* 11 February 1897, noted that "the brilliant editor of a 'society' weekly in this city solemnly adjured all prospective guests to study carefully the characters of the personages they intended to represent, so that they might render their conversation historically accurate, and avoid anachronisms."

23. Burrows and Wallace, 1171; *New York Times,* 25 January 1897; *New York Tribune,* 11 February 1897; *New York Times,* 31 January 1897; Godkin, 495.

24. Soissons, 91.

25. Guy de Soissons thought it "mean" of American millionaires to "erect high fences around their palaces, in order not to give anyone a chance to see the house, and, by looking, to develop his good taste. If, by chance, the gate were open and anybody approaches it to look at the house, the porter, rough as his master, will drive away the indiscreet transgressor; and so, the man who has no mind of his own, seeing the roughness of a millionaire, will, in his turn, think that it is the proper thing to be rough" (94–95).

26. Welling, 271.

27. *New York Herald,* 24 December 1893. Although Ward McAllister, toady and social arbiter, did not live to see the Bradley Martin ball, he had earlier defended such expenditures in *Society as I Have Found It* (1890): "The mistake made by the world at large is that fashionable people are selfish, frivolous, and indifferent to the welfare of their fellow-creatures; all of which is a popular error, arising simply from a want of knowledge of the true state of things. The elegancies of fashionable life nourish and benefit art and artists; they cause the expenditure of money and its distribution; and they really prevent our people and country from settling down into a humdrum rut and becoming merely a money-making and money-saving people, with nothing to brighten up and enliven life; they foster all the fine arts; but for fashion what would become of them? They bring to the front merit of every kind; seek it in the remotest corners, where it modestly shrinks from observation, and force it into notice" (160–161).

28. Bates, I, 250; Burne-Jones, 88–89; Richard Wells, 402–408, 458. The great shark Jay Gould loved orchids and grew them in his vast Moorish-style greenhouse at Lyndhurst; at his burial in 1892, however, even a huge cross made of orchids was not enough to distract the family from the sight of workmen painstakingly sealing the casket against bodysnatchers with hot metal, one spoonful at a time. See Klein, *The Life and Legend of Jay Gould,* 212, 216, 415.

29. Some ballgoers who attempted to dance experienced intense difficulty with their costumes. Several men became entangled in their swords, and a woman's powdered coiffure "exploded, and the powder filled the eyes and nose of the man, who therefore sank down upon the floor and sneezed and sneezed and sneezed. What added to his embarrassment was the fact of his not being able to remember off-hand in what section of his clothing a courtier of the Seventeenth century kept his handkerchief" (*New York Sun,* 11 February 1897).

30. On 2 September 1893, Mayor Thomas Francis Gilroy announced that New York City was the richest city on earth, "the best practical answer to

those critics who maintain that the affairs of the city are unwisely or dishonestly administered." The *Herald* was prone to such announcements as "All you need is money," "Money is king," and "In this city money marches before everything else" (2 September 1893, 19 November 1893, 1 April 1894).

31. *New York Herald* and *New York Times*, 14 April 1899, 17 May 1899, 28 May 1899, 6 June 1900, 5 April 1901, 17 April 1901.

32. On 17 April 1893, the *Herald* explained the altered view of the social rich: "In 1876 society journalism was in its infancy. The daily doings of that elect throng which in those days was even less than the modern '400' was almost scorned by the daily papers. Since that time everything has changed. Social doings are eagerly looked to by those of high and low estate, and though there may be a show of reluctance to have such items paraded before the great public, there is a great amount of satisfaction among those who are thus frequently mentioned."

CHAPTER 2

Private Showing

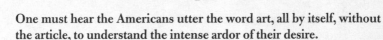

One must hear the Americans utter the word art, all by itself, without the article, to understand the intense ardor of their desire.
—Paul Bourget, *Outre-Mer,* 1895

IN THE CLOSING DECADES of the nineteenth century, Americans across the country spoke a now forgotten language of *art.* Within its lexicon, an American, whether male or female, could not only be *artistic* but also have art instincts, art ideas, and art feeling. He or she could be an art personality or an art enthusiast, have art connections, follow art opinion, live in an artistic house, fill it with art furniture, belong to the artistic class, wear artistic clothing, and cultivate an artistic goatee. One could be an art worker, lead an art life (though this desirable situation was mostly available in France and Italy), create an art atmosphere, live in a vital art element, and seek the art spirit. Late-nineteenth-century America allotted considerable physical space to art, and then crowded every surface of that space with all it could hold in the way of paintings, statuary, fabrics, rugs, *objets d'art,* color, and pattern. Because city space was not only expensive but also primarily horizontal, allotting so much space to goods meant less space allotted to the bodies that had to live in the spaces or at least negotiate a path through them. When expanding artistic desires came up against limited physi-

cal space, busy and ingenious nineteenth-century people found a solution: they decorated, and thus elevated to *art,* objects previously
thought of as utilitarian, and they maneuvered sites, behaviors, and
people into an unlimited virtual space labeled *artistic.* There were
advantages. Once lodged in *artistic* space, a person could experience
a considerable increase in personal freedom and even carry special
temperamental cachet: he or she could be nervous, sensitive, passion-
ate, and magnetic, rightly self-concerned and easily annoyed by lesser
beings. An artistic life, in the terms that the nineteenth century was
willing to assign it, came to seem "somehow bigger and subtler and
darker," as Theodore Dreiser put it in *The "Genius,"* than other orders
of living.

If to be artistic was just the thing, to actually work as an artist in
America was quite another. So much human capital invested in *being
artistic* left little to invest in the artist, and so, while *being artistic*
gained cultural validation, being an artist lost it at a parallel rate.
Excluded from the art fervor were American-born artists working in
America, a number estimated to be about thirty thousand in the
1890s; in Manhattan there were perhaps three hundred artists, not
including students. The prestige of American artists, however, was so
low that no collector could raise his or her status by buying American
art; most big buyers in the art market refused to transact with any
American artist.[1] At the World's Columbian Exposition in 1893, the
Chicago Tribune reported that agents were on hand to sell European
artwork, but "no provision seems to have been made for the sale of
works of art by American artists. It may be that the possibility that an
American citizen should purchase a work of art by a fellow-country-
man is so slight that any preparation for such an unlooked-for event
was considered unnecessary." At the Hotel del Monte in San Francisco, James Nelson Fraser watched the "crowd of millionaires" staying at the hotel ignore its sale gallery of American art, "where so much
talent is evinced and so much labor and so many hopes are buried.
The prices of the pictures are piteously low, and nobody buys them.
Thus it is all over America. Not talent was wanting nor interest in
drawing, but patrons. I have read that over a million pounds worth of
pictures are bought in Europe every year for America. In the past

much of this money has been wasted; it is stated that twenty thousand forged Corots have passed through the American customs."[2]

The very category of "American artist" was troubled by American ideas about business and work that did not apply to European artists. Lodged in Europe, an artist of eccentric habits could maintain a romantic aura, and, since masculinity was constructed very differently in Europe, he could perform such services as accompanying his client to the House of Worth, where he might help to design the opera cloak she would wear in her portrait. In contrast, an artist living in America seemed, in the climate of the time, unpleasantly eccentric, a person of dubious habits, no business sense, and a murky connection with the world of work. In 1904, Philip Burne-Jones noted "how very funnily people in America, otherwise honorable men, will behave to artists, with whom they seem to feel safe to play tricks which would never be tolerated in ordinary business relations." But there was, at one and the same time, an insistence that art and artists should connect with business values and should be discussable in business terms, even though Americans were refusing to do business with them. The *New York Herald* could, in one moment, divide business from art and announce that "behind the footlights" of the Metropolitan Opera, "it is business there, not art," and in the next moment claim that "bad sculpture" was "rampant in this city" and had to be removed before it damaged the city's "commercial prosperity."[3]

Businessmen dreamed of profits from an art scene that they refused to support. In 1902, Brook Adams, speaking at a dinner of the National Art Club in Manhattan, told his audience that "it was a sound commercial instinct which led a people to devote its talents and energy to the creation of a great and original art." Such an art, according to Adams, could "attract money-spenders from all over the world." If New York itself became a "thing of beauty," he insisted, it would "save for her shop-keepers the one hundred million per annum which Americans now pay into the pockets of the shrewd and businesslike French and Italians." *Architectural Record*, which reported Adams's after-dinner speech, was not convinced: "What we need in America, in order to have an art that pays a good 10% on the investment, is an inherited ancestral art legacy, and our ancestors were, I am afraid, entirely callous to our opportunities of profiting from their artistic work." Art, concluded the *Record,* in a voice of increasingly bitter irony, "is a worthless investment." Meanwhile, an evangelist such as Dwight Moody dreamed of tossing all the "vile pictures" onto an artistic bonfire: "I am tired," he told a revival meeting in Carnegie Hall in January of 1897, "of what they call art. The time has come to save our children. The Nation is decaying. We are going the way of Babylon, Nineveh, and Rome, and it is going down under the polite name of art."[4]

Socially speaking, rich Americans rebuilt old barriers between artists and society, barriers that had crumbled in Europe over a century earlier. They might accept an architect or a decorator—at the time often the same person—if he or she could demonstrate social status or considerable financial success, and they might respect an American-born artist who had conducted his entire career abroad—John Singer Sargent, James Abbott McNeill Whistler, Henry James. They lionized the foreign-born novelist and fought for the services of the visiting foreign-born painter. The French Saloniste Benjamin-Constant criticized Americans who studied painting in Paris as being well prepared but "filled with the false idea that their aim should be to create new works . . . to branch out in another direction." Even those American artists who lived abroad, carefully imitated French Salon styles, and met Salon standards were hard pressed to find an American market, and in general could get for their pictures only a quarter of what a French Saloniste could command.[5]

From another angle, American artists were criticized by one George Burroughs Torrey, a European-trained American who had found success in painting society women as "goddesses." Seated in his "flawless Louis XV drawing room" in New York, Torrey explained to newspaper reporters how, upon his return from Europe, he had built his art career in America.

> I did not do as the artist usually does. I did not go to a studio building and take a room where I would see only artists. I went to stay at the Waldorf, then newly opened, although I had very little money to do it on. But there I met statesmen,

bankers, men of affairs from every part of the country, and that stay of four years in the Waldorf was worth thousands and thousands to me. Here in America, a man will go to an artist and say, "I like your pictures, but they are in too high a key for my house, which is in dark wood with tapestries. Won't you paint me a low-toned picture that will harmonize with the surroundings?" "Blank your house," the artist says, instead of painting what the man wants.

J. G. Brown, fabulously successful painter of multiple Algeresque newsboy pictures, took the same line when he explained to the *Herald* why American artists failed:

> My experience is that most artistic failures—failure to win financial rewards as well as artistic honors—occur because the young and needy painter regards art solely from an artistic standpoint. He is apt to injure rather than advance art by going off mooning after the ideal. My advice to young painters who want to make their way to life's comforts, as well as art's honors, is to remember that commerce and art touch elbows.[6]

Henry Alexander, a Munich-trained American painter, might have modeled for Torrey's and Brown's ideas of a failure. On May 15, 1894, unable to pay his rent, he was locked out of his room in the Tenth Street Studio Building. Alexander spent his last day on earth wandering Manhattan, carrying his finished painting of an interior scene in the Hebrew Orphan Asylum. William Merritt Chase saw him trying to sell the picture to a Fifth Avenue art dealer. When he failed to make a sale, Alexander engaged a room in the Oriental Hotel at Broadway and Thirty-ninth Street and drank carbolic acid.[7]

The lonesome Henry Alexander, toting his possibly grim little picture from one dealer to the next, was up against one of the glossiest and longest-running blockbuster shows of all time: the Paris Salon, the annual exhibition in Paris of thousands of academic paintings stacked frame to frame up the walls of the many rooms of the Salon. Hundreds of those paintings were further boosted in value by prizes, ribbons, and citations. Salon art was certified as unquestionably "the best"; French pictures were the best. In cities across America, the rich tumbled over each other to assemble private collections of art validated by display in the Paris Salon and to build their own private blockbuster shows in their own private galleries. Americans hung their personal shows of Salon art just as they had been hung in Paris: thick gilded frame pressing against thick gilded frame, stacked upward from the dado to the ceiling and then, if need be, pressing downward from the dado to the floor. Some collectors acted as their own docents, reciting to selected visitors the names of the painters, the titles of the paintings, and often the prices paid for the artgoods.

The art historian John Van Dyke estimated that all the public and semipublic collections in New York "put together do not equal the quantities of fine art in the New York private houses. There are hundreds of galleries of pictures, with bronzes, fabrics, and furniture, in individual hands." When Van Dyke offered the hopeful guess that private art holdings "did educational service in a quiet way among coteries of friends," he admitted that these much publicized collections were generally unseen. Most late-nineteenth-century artlovers were hungry for images but belonged to no such coterie of friends of the rich. Even those unaware that they were art lovers would have lined up to buy tickets to such a show, because in America any blockbuster show was and is a no-risk proposition: viewers know before they enter that it is good and that it does not require their critical assessment.[8]

A few Salon-art collectors issued invitations or cards of admission to their private galleries, but only to certified and special persons. One such select visitor was Winefred, Lady Howard of Glossop, the kind of British personage who, in the American climate of the 1890s, was certified by her title and knew it: on her arrival in New York she paused to look about for the mob of interviewers who would surely be wanting her opinion of a country she had not yet seen. Lady Howard was, however, a serious and devoted artlooker. Undaunted by the three-day blizzard of February 1896, she set out from the New Netherland at Fifth Avenue and Fifty-ninth Street to see art. At the Metropolitan Museum she saw the goods that had already made their way from A. T. Stewart's collection to the Met, including Meissonier's *Fried-*

land and Rosa Bonheur's "magnificent *Horse Fair,* a glorious piece of color, form and movement, and light and shade. You seem to hear the very ring of the horses' hoofs as they trot to the show!" From there, however, Lady Howard went on to see "art collections in several private houses," shows to which most Americans could not gain admission. On February 8, because "no cabs were to be had" in the suffocating blizzard, she struggled on foot, battered by "fearful gusts," to honor her engagement to see William H. Vanderbilt's "beautiful art collection." Though she did not say so, she could not postpone her visit because Vanderbilt opened his gallery on Thursdays only, between eleven and four, and then only to persons positioned to obtain "cards of invitation." And though she did not say so, the indispensable *Artistic Houses* makes it clear that she did not enter through the front door: Vanderbilt had had constructed on Fifty-first Street "for the convenience of the public . . . a special entrance to his picture-gallery," done all in red and gold, and connecting to "a dressing-room, with toilet conveniences for visitors," carpeted and finished in mahogany—thus giving visitors no excuse to ask for entry to any other part of the house. A particular drawback of the crowded urban private collection emerged when Lady Howard judged the picture gallery to be "scarcely light enough." Nonetheless she was, as she wrote, "well rewarded by the art-feast I found there: nine or ten admirable Meissoniers, three of Millet's most poetical and lovely works, now priceless, and for which he himself received such miserable pittances, and many other beautiful paintings and objects of art."[9]

Urban artlovers generally had their chance to see a large private collection only at its sale, at that moment of dispersal when it ceased to be a collection. Some collections were so large that they had never been displayed before they were sold: George I. Seney's third collection of 317 pictures was so extensive that, according to the *Herald,* "the pictures were never all hung, and were never even all together in the same building." Such private collections were closely held but dramatically unstable, and throughout the closing two decades of the century, pictures bounced from hand to hand as the fortunes that had bought them rose and fell. By 1897, private collections were recirculating with such speed that the *New York Times* deemed it "necessary" that the "art lover who wishes to keep in touch with, or abreast of, art progress and life in America, pay at least a weekly visit to the best-known galleries and art stores." At auctions, more and more pictures were brought before large audiences whose interest seemed to be flagging and whose bidding was often described as spiritless. In the chaos of exchange, values became confused, and strange bubbles arose: for a brief time, the work of Ludwig Knaus—genre pictures with titles such as *Excuse Me, Sir* and *Helping Hands*—brought the highest prices of the evening while a solitary Monet was recaptured by the French for $550 and a lone Ingres went out the door for $65.[10]

In 1887, John Tod, a Scots writer who liked to quote Robby Burns's lines "A chiel's amang ye takin' notes,/An' faith he'll prent them," left a rare record when he happened into the sale of the late A. T. Stewart's collection of paintings, sculpture, ceramics, plates, and bric-à-brac. Few Americans had ever glimpsed the private gallery in Stewart's marble mansion at Fifth Avenue and Thirty-fourth Street. Tod had a special interest because Stewart had left many relatives behind when he emigrated to America, and "some worthy Scottish folks in the southern counties expected a fair slice" of the Stewart estate. They got nothing, for, as Tod put it, "the slip between the cup and the lip was a complete spill." Tod was also interested in "observing the kind of pictures with which a busy merchant and clever man of means and money surrounded himself. Most of the pictures were bright and suggestive, and told their own story without the aid of a catalogue. Of old masters of the dingy type there were none—color, sparkle, and bright effect were the features; what, in Europe, are considered masterpieces, and deservedly so, although dingy, smoke-begrimed, and cracking, were wanting." Although Tod dismissed Meissonier's *Friedland 1807* as "a battle piece of the palmy days of Napoleon I," he admired the work of such other big Paris Salon names as Bouguereau, Troyon, and Gérôme. There were 220 pictures in the Stewart collection, and the sale of all his goods occupied three evenings at Chickering Hall, a two-thousand-seat auditorium on Fifth Avenue at Eighteenth Street. The hall was packed for the occasion.

John Tod arrived early for a good seat. He watched as Jay Gould and William H. Vanderbilt were ushered to the front row, and he noted "young men ranged at short distances in the aisles, to catch bids that would have been lost in the crowd without such help. Mr. Thomas Kirby, a man in his prime, stood at a small desk; behind him was a platform, half concealed by deep, large red curtains. Mr. Kirby spoke incessantly, repeating the last bid. Bids were bawled out by the assistants in the aisles, even from far back in the gallery." The red curtains swept back to reveal each painting, and then, when it had been sold, swept forward again while it was replaced with another. The enthusiastic crowd greeted several pictures—and any good bid—with applause, "and a picture of George Washington by Gilbert Stuart brought down the house." A. T. Stewart's collection was decidedly mixed: some very large Salon paintings brought equally large prices, which Tod enjoyed converting into dollars per square inch of canvas, but such "Old Masters" as Stewart's Titian, his Murillos, and his Rembrandt struck Tod as "in art phrase, hardly dry—the brightness of the coloring being suspicious, and buyers seemed of the same opinion. Many pictures fell far short of what had been originally paid for them." Furthermore, there was nary a bid for Stewart's edition of Hiram Powers's *Greek Slave*—a nude from an American hand—and it was withdrawn. "Great is art in America," wrote Tod, "and liberally is it patronized. Greater still is American propriety."[11]

Art poured into the United States from Europe: pictures, marbles, tapestries, porcelain, furniture, medals, plate, and rugs. Foreign visitors at the turn of the last century were awed when they took the measure of the artgoods that had emigrated to America since 1876; Americans, they said, had "laid violent hands" on European culture. The British Charles Whibley called the contents of private collections and museums "the despair and admiration of the world," but to German art authority Paul Clemen the export flood was an "American danger which has assumed terrifying dimensions and is earnestly threatening European art treasures further." To European travelers, American collections spelled loss—lost national property, lost cultural property, and even lost personal property, especially when, on entering a private gallery, they encountered a special favorite now in the hands of a rich American. In 1894, visiting James J. Hill's house in St. Paul, Minnesota, Paul Bourget came upon a familiar Delacroix.

> It was a view of the coast of Morocco, before which I stood long, as in a dream. I saw this canvas years ago. I have sought for it since in hundreds of public and private museums, finding no book which could inform me who was its present possessor, and I find it here! What ground has this canvas covered between the painter's studio and the gallery of a millionaire of the Western frontier! All the glory of France! What sentiment impels these wealthy speculators to gather into their own homes art treasures most foreign to all that has been the business and passion of their whole life?[12]

In the twilight of the nineteenth century, during exactly the same decades when American museums were coming into being all across the country, the appeal of the private gallery was at its zenith. Art collections appeared not only in the private galleries of the rich and the aspiring, but also in bars, in hotels, in business offices, and in the art-units and art-corners of the average private house, apartment, or boardinghouse room. Human beings posing as artworks appeared in theaters as *living pictures* and *glyptoramas,* and before private audiences as *tableaux vivants.* For the most private viewing of all, art appeared in bulky peepshow books with the words *Gems* and *Masterpieces* in their titles. From the mid-1870s to just exactly 1900, many thousands of the Salon paintings favored by the rich entered America; these often immense canvases were stocked with nearly life-size photorealistically rendered figures of humans and animals in scenes that bore no visual connection to life or landscape, urban or rural, in America. After two decades of rapid movement from hand to hand, the paintings were unloaded on the new museums of the time. There, during the first quarter of the twentieth century, they moved from gallery to stairwell to storage to oblivion. Many were eventually deaccessioned and came to decorate hotel lobbies, function as theater backdrops, and form the side panels of circus wagons.

For a time, until Prohibition swept them out of sight, many Salon pictures hung in bars. On January 25, 1920, after one week of life in a dry America, a wag writing for the *New York Sun* asked:

> What will become of the art treasures of saloons? There must have been, before the 17th of this fateful month, twenty thousand bars which had one or more paintings to which the owner pointed with pride and which the patrons surveyed with mixed emotions—a decorous nude or two, a Holland scene, a Parisian shop girl, a flock of bacchantes reeling across a green lawn, just plain "Spring," or the conventional woman prone upon the sand. The proprietor always knew the name of the painter. "That's by Bazinkus," he would explain, "and I've had some big offers for it from prominent business men." But now that there are no more cocktails it is likely that prominent business men will stop making fervent attempts to gain possession of these pictures. Any object of art looked good after the seventh Martini. Must all these alluring canvases go to the attic?

After considering the possibilities of a "rum museum" stocked with "relics of mahogany, mirrors, brass, curious utensils, and a few hundred of the best saloon paintings," he hit on a better solution, the Hackensack meadows:

> They are a sad place, particularly in winter when trains are stalled. The advertising signs are dreary things to look at while the engineer readjusts the carburetor of his locomotive. A first class saloon oil painting stationed every five rods across the marsh and lighted at night would make the home-going of the commuter a brighter and more human journey. Governors Edwards and Smith might appoint an interstate hanging committee to look into this.

Meanwhile, uptown at the Metropolitan Museum of Art, trustees contemplating their own wealth of "reeling bacchantes" entertained their own wag who proposed to "store them in a non-fireproof building, insure them at full value, and leave the rest to Providence."[13]

For a time, however, and across the duration of the episode in American taste under examination here, French Salon art ruled in America. It did so under the covers of American devotion to art, admiration of big price tags, ambivalence toward the female body, and worship of all things French.

France was a magic word in America at the end of the nineteenth century. For twenty years at least, those recognized as artists in America were European and most especially French; in painting, sculpture, and all things decorative, France held sway. The Frenchifying of New York drew mixed reaction from foreign observers: while Henry Vivian in 1878 was impressed by the power of the illusion and by the way New York "exerted Parisian influence throughout the whole domain of the Union," Samuel Day in 1880 judged "the attempt of the fashionable New Yorkers to imitate the French in everything to be pitiable if it were not despicable, a rude burlesque of the reality. Nature has not gifted them with the grace or the genius of the admired foreigners." East Coast urban Americans nonetheless kept their eyes fixed on France: streams of orders for paintings, sculpture, clothing, jewelry, *objets d'art,* and furniture went to France, and decaying French estates were ransacked for tapestries, doors, and ceilings that could be bought and reinstalled in American mansions. "Everything French," wrote a *Herald* reporter in 1893, "appears to be enjoying a boom in New York at present. If this thing continues, the customs and language of gay Paree will drive the English faddists out of the market. Nowadays nothing is considered worth seeing or hearing except what is French."[14]

In that same year, when the French novelist Paul Bourget toured America, his every utterance was reported and scrutinized for suggestions of attitude; the *New York Herald* struggled not to lose faith in his French superiority even when he pronounced the hated city of Chicago to be a miracle. "It is possible," mourned the *Herald,* "that the barn-like interiors of some of the buildings were a disappointment to him, but he would not say so." The very word *French* functioned as both protective cover and signal. When Earl Lind visited Paresis Hall on Fourth Avenue south of Fourteenth Street in 1895, he met a "little club" of bisexuals united for "defense against the world's bitter perse-

cution." They called themselves the Cercle Hermaphroditos and cus-
tomarily addressed each other in French. Specialized sexual practices
were coded as *French treatments* and *French love*. When large groups
of big-city Americans wanted to behave badly, engage in high kicking
(the code of the time for displays of women's genitals), fornicate in
public, and drink themselves into a stupor, they called the occasion
French.[15]

New York City's annual French Ball had a long run—from 1866 to
1901—through several changes of venue; in the 1890s it was held at
Madison Square Garden. The event had no connection whatever with
New York's French community, estimated at about fifty thousand per-
sons, and in fact no connection with France beyond using it as a cover.
The ball was sponsored by the Cercle Français de l'Harmonie, a society that the Lexow Committee[16] called "a standing disgrace to New
York" but whose officers considered it to be "just as honorable a society as any other society in New York." Whether honorable or not, the
French Ball had, at least until 1894, the protection of numerous high-
ranking New York City police officers who attended—including
Superintendent Thomas Byrnes himself. Agents for antivice and anti-
crime societies also swelled the crowd, eager to study at close range
behaviors that might require suppression. What Timothy Gilfoyle in
City of Eros calls the "confusing sexual boundaries of New York" were
drawn in very specific ways by the French Ball. Men of all ages, many
of them upper or upper-middle class, put on evening dress, left
mother, wife, and fiancée at home, and crowded into the Garden to
party all night with young women of a distinctly lower class, many of
them prostitutes. No clearer example is available of who had which
liberties: men with money, regardless of age, looks, or class, could, for
the length of the ball, have their chance at innumerable young women
who had only their looks and their bodies to offer.[17]

On February 11, 1893, the New York newspapers went to the
French Ball. For hours before the dancing began at midnight, "carriages blockaded the streets for half a mile around the Garden" and
"everybody fought for standing room at the entrances. Judges and ballet girls, matrons and painted sirens, bunco steerers and clergymen
churned and hustled like a mob at the races. There was not any notice-

able anxiety on the part of reputable citizens to disclose their identity to the crowd. As a rule men did not care to answer to their names." *Herald* reporters described the scene as "a blizzard of passion" and "a riot of debauchery," with sex acts being committed in the Garden's family boxes, comatose girls, old men "crazy with wine," and the floor littered with "trophies of shame." The *Times* headlined "Jezebel Holds Carnival—an orgy winds up the big French ball." In the ensuing confusion and uproar over the newspaper reports, anything French briefly became potentially orgiastic, and the police, who had looked away from the French Ball, reasserted themselves by cracking down on a rather staid family ball thrown annually by the Mardi Gras Association and sponsored in part by the French consul, at the Lenox Lyceum on West Fifty-ninth Street. At midnight on February 20, 1893, the police raided the Mardi Gras ball and turned New York's French colony out into the street, leaving fifteen hundred French dinners untasted.[18]

In 1894, the *Herald* peeked out at the French Ball from behind the ministerial person of the Rev. Madison C. Peters of Bloomingdale Reformed Church at Broadway and Sixty-eighth Street. Peters was lured into reconnoitering the situation as a chance to denounce sin. Instead, he denounced men. It was Peters's first public ball, and after describing a scene of "disgusting nastiness and bacchanalian debauch," he warmed to his subject and took on half of the guests:

> I never saw a finer, more respectable looking crowd of men anywhere. The old men, the gray headed, and the bald were out in full force, married men, fathers, scores of them. Society is full of men—excuse the mistake—filthy animals devoid of heart or conscience; incestuous brutes, luxurious in inherited wealth; men of years and gray hair who often boast of the histories of their amours. New York society is full of men whose lives are so notoriously corrupt that their very touch is pollution and their embrace social and moral death. I have no plea for her who is led to sin, but I want a change in public sentiment that will lift four points of the guilt from the head of the fallen woman and hurl it in withering condemnation upon the man who opened for her the gate to sin.[19]

Madison Peters had no effect. In the following year, the French Ball, "New York's kaleidoscope of wanton wickedness," was more chaotic than ever before. Ballgoers destroyed the floral decorations and smashed the flowerpots into smithereens, tore off their own and others' clothes, and cut each other up with broken champagne bottles. Cross-dressed women fondled each other, and men—a group that included "high officials and ex-rulers of the metropolis, grave Commissioners of Education and police justices, business men and leaders in half a dozen learned professions"—stood on their heads and then "encouraged their female companions" to do the same. Dancers stepped over comatose ballgoers lying wherever they had passed out. In general, judged the *Herald,* "there was no feeling of shame."[20]

Unlike the homegrown French Ball, French Salon art came into the United States strengthened by real French credentials, validated by selection committees and official exhibitions and prizes—it was, in France, *l'art officiel.* Salon art was not, however, bolstered by European critical acclaim: numerous prominent European art critics had been groaning over it as iterative and dead for two decades before it arrived in the United States. Americans who could visit neither the Salon itself, nor the studios of French artists, nor the private collections of the very rich could nonetheless read about Salon art in art periodicals, and they could view it through the eyes of Earl Shinn. Immediately after the Philadelphia Centennial Exposition in 1876, when Paris Salon art began to pour into America, Shinn set out on an odyssey that eventually took him through 177 private collections in every major American city except Pittsburgh. Shinn examined thousands of "modern" pictures, 2,289 of them by 411 French artists, 70 of that total by Adolphe-William Bouguereau, American collectors' top Salon favorite. Shinn chronicled his search in three illustrated volumes titled *Art Treasures of America.* Despite Shinn's agonized ambivalence toward his subject, his massive effort operated to spread the taste for Salon art across the country. The effect of *Art Treasures* was further bolstered by *Artistic Houses,* an extremely influential 1883 series that offered photographs and descriptions of the décor and art in ninety-six domiciles owned by the rich and important. Over the next twenty years, innumerable traveling exhibitions, volumes reproducing "French Masterpieces," and newspaper and periodical coverage of art further enhanced the value and "beauty" of Salon art.[21]

This extraordinary quarter-century episode in American taste was tightly enclosed in its time framework. Its collapse in value around 1900 was complete. As an American consumption phenomenon, Salon art was a monster, supported by rich collectors who issued commissions, whipped fresh examples off the easels in French studios, and circulated favorites among themselves in the auction rooms. No amount of biographical detail about individual collectors, artists, and dealers can illuminate this consumption extravaganza; it is not about individuals but about social agreement and taste culture, about a group effort to transport an enormous heap of artgoods across cultures, from public display in France to the private gallery in America, from an old culture to a new one, from a teeming art world to one they thought quite bare. Collectors of Salon art were not driven by desire for the one-of-a-kind item; such desirable Salon paintings as Alexandre Cabanel's *Venus*, now in the Metropolitan Museum of Art, were done in multiples, and American collectors did not mind. Moreover, across their relatively brief history in America, collections of Salon art moved toward complete standardization. No collector claimed that he or she viewed Salon art as an investment, or spoke of fashion, or talked about simple pleasures for simple persons. Except perhaps to identify the artist and quote the price, these intense collectors said very little about the objects of their desire. Salon art, which was one type of cultural event in France, became a very different type of cultural event in America.[22]

French Salon paintings are best distinguished one from the next not by style and not by individual artist but by subject matter, and American collectors bought so much Salon art that they influenced that subject matter. By the end of the century, French Salonistes were in fact producing for the American market; an episode that began with Americans selecting among Salon subjects ended with Salon painters rendering the subjects American buyers liked best. On the favored-subject list were pictures of a European peasantry and humble genre scenes of European life; classico-mythological narrative pictures;

"church paintings" of religious subjects and persons engaged in religious practice; "cattle paintings," very large renderings of cows, horses, sheep, ducks, geese, and chickens—but no pigs—that, in the rapid sale exchanges of the 1880s and 1890s, drew some of the highest prices; Orientalist art, especially Algerian and Egyptian "harem scenes"; and battle pieces, these frequently involving Napoleon. The catalogues and indices of *Art Treasures in America* suggest that individual collectors specialized in such subjects as women—widows, babies, young girls, women getting dressed and undressed, women bathing, and women sewing and knitting; wine-tasting, cigar-smoking, and pipe-smoking; food, with cherries and chocolate foremost; musical performers; and domestic pets, especially kittens and parrots. Across all the collections the two most frequent narrative subjects were, first, Cupid and Psyche, represented as anything from infants to mature adults; and second, Marguerite of Goethe's *Faust*—Marguerite weeping, Marguerite waiting, Marguerite in prison, Marguerite in paradise.

Standardization of Salon art collections in America occurred early in the phenomenon: the art collections of the ninety six houses visited and described for *Artistic Houses* in the early 1880s are noticed again and again for the same painters. In 1902, W. G. Constable, already looking back on these collections as a vanished phenomenon, described them as "all determinedly contemporary, rarely including Old Masters, and only a limited number of works by nineteenth-century painters already dead. The staple of the collections was the work of the leading French academic painters of the day, notably Bouguereau, Gérôme, Meissonier, Cabanel, Lefebvre, Rosa Bonheur and her brother Auguste, Detaille, Berne-Bellecour, and de Neuville." German art in these collections covered the same range of subjects as did the French, and other favorites such as Fortuny and Zamacois the Spaniards, Alfred Stevens the Belgian, and Munkácsy the Hungarian were "all influenced from Paris." Constable also pointed out that "in all these collections British paintings were hardly represented; American artists came off little better."[23]

Satisfying American demand for pictures of a European peasantry kept Adolphe-William Bouguereau busy at his easel. American collectors lavishly mourned the early death of Jules Bastien-Lepage, who united his peasantry with religious and mystical themes; and the staggering popularity of François Millet in the 1890s lodged peasant pictures, to adopt Mary Douglas's term for such consumption phenomena, at the "hot center" of the formulaic art collections of the time. Certainly when a rich American purchaser of European-peasant pictures entered his private gallery, he was no longer merely someone who chased dollars. He was a collector who desired the company of some very specific art. The question is what he saw when he looked, and whether the world in which he had become rich still awaited him in the shadows of his gallery, covertly, behind the glossy surfaces of his Salon pictures.

The lure of the peasant pictures lies in all that they are not. In many peasant pictures, Breton peasants perform a style of manual farm labor that, in an America where the self-scouring steel plow had been introduced in 1836, few had ever seen. Furthermore, the peasants work on a stump-free landscape that bears no resemblance to the rural America of the time. The peasant women are engaged in fieldwork, a gender situation utterly unacceptable to Americans in the nineteenth century; some are performing stoop labor, a type of work Americans had historically refused. The peasants are usually shoeless, while every American farmer wore boots; the women wear peasant blouses and little peasant caps, neither garment ever seen in rural America. The peasants are unthreatening noncompetitors, probably illiterate, looking about in a vacuous or possibly contented way, unlikely to contest their control by invisible landholders. Salon peasant pictures rarely suggested the fact that in late-nineteenth-century France the social position of the peasant was a very hot political topic. Indeed, America's favorite, Adolphe-William Bouguereau, skirted the political and social implications of his subject by obliterating all working context from his peasant pictures. Bouguereau's barefoot peasant girls and women stand rather idly by a well, or lounge about telling each other "secrets," or hold a needle in one hand while staring off into the airscape.

Transported to America, a country where some women suffered from forced disoccupation while the rest suffered from the seventy-two-hour workweek, the peasant pictures ignore American working conditions for women; the pictures honor a type of work American women did not do and at the same time deny that women do any work whatever. The raised heads and unplied needles of Bouguereau's peasant girls obliterate the American reality of women and girls who labored, heads bent over their sewing, twelve or more hours a day in New York's tenement manufactories. Every Salon peasant abides outdoors; no one is imprisoned in an airless sweatshop or an unventilated tenement room. In 1888, Clarence Cook, the American arbiter of taste, read the peasant pictures as playing to class attitudes:

> Since one of the things that rich and fashionable people take pleasure in, is the knowledge that there are others in the world who are not as well off as themselves, Bouguereau has provided for his admirers an ample supply of pretty beggar-children, young peasants, mothers of the picturesque poor, who are in truth people of the upper class, or seem to be such, with beautiful faces, fine heads of hair, rich eyes, chiselled lips, ivory skins, hands that never wore gloves, feet that never wore shoes, and all in a state of immaculate cleanliness, and in garments where, if a patch or two is to be seen, it is accepted as mere symbolism and offends nobody.[24]

The iconography of peasant pictures delivers some messages only to those equipped to read them. The peasant blouse is one such message. Lightweight muslin, cut low or cut across from shoulder to shoulder, gathered in below the breasts with a laced bodice, the peasant blouse was a hot piece of nineteenth-century erotica: a garment of both emphasis and concealment, drooping over one shoulder, almost falling off, a thin covering over an uncorseted and possibly available body. In America, any corresponding garment would have been underwear—a chemise or a corset cover. The Salon-art peasant blouse sent sexual signals whether it was worn by a young woman or by one of the preadolescent girls who were of consuming interest to late-nineteenth-century American males. In 1901, when Stanford White took sixteen-year-old Evelyn Nesbit, the object of his desire, to be photographed, one result was a stunning photo of Nesbit in peasant blouse, seated on the floor and leaning suggestively toward the camera. The peepshow "masterpiece" books of the time are dominated by pictures of women losing their peasant blouses and looking hotly out of the canvas. The message about peasant bodies went out to women also: in April of 1894, in New York, a Dr. Kelley delivered to an all-woman audience a lecture on "the perfectly formed woman," illustrated with stereopticon views of nude women. Dr. Kelley claimed that the most perfect woman he had ever seen—though he had no photo of her—was "a peasant, who made her living by carrying heavy weights upon her head."[25]

The sexual content of French Salon peasant pictures is further amplified by their loving attention to the feet of their subjects. Bouguereau in particular was a foot specialist, and commentators who couldn't find a thing to discuss in a Bouguereau picture relied on admiration of his pretty way with toes. It was a premise peculiar to 1890s America that the peasant foot was beautiful, never deformed by the shoe—the shoe that was being manufactured in city sweatshops. In foot fantasy, a person who had gone barefoot for a lifetime would have clean, beautiful, and perfect feet—just the feet that Bouguereau painted. In 1890, Richard Wells, in *Manners, Culture, and Dress of the Best American Society*, rhapsodized over "the shapely feet of Italian and Spanish peasants, which never had known the bondage of a shoe," and then swung into detailed description of "a modern, fashionable" woman's hideously deformed feet and equally lingering description of Wells's own notion of the beautiful foot. Although Wells, like Dr. Kelley, sent his messages to women only, beautiful-foot messages received the full support of American men. If the French critic René Ménard scoffed at Bouguereau for "the naked feet he gives to his peasant-women that seem to be made rather for elegant boots than for rude sabots," the American artist Carroll Beckwith was "reminded of the masterly study which must have preceded such excellence" in the drawing of feet, and Clarence Cook, who despised Bouguereau and called him "an industrious plodder," nonetheless admitted that "he draws hands and feet irreproachably; in fact, he is recognized as a master in this field." Earl Shinn lingered over Cabanel's *Venus*, now in the Metropolitan Museum of Art, and panted that "one nonsensical little great-toe of hers curls up the reverse way—

a blown rose-petal, to be kissed, maybe." Of Cabanel's *Shulamite*, Shinn wrote that "the gauzes fall from the bosom of the sacred odalisque, and the rose-petal nudity of the feet tell of the warm ennui of sun-baked lands." The bare foot has more than one sexual possibility: "If, in imagination," as Tracy C. Davis points out, "the feet were bare, so were the legs, and if the legs were bare the potential revelation was unlimited."[26]

The foot seems to have been a more open secret than was the peasant blouse. Feet were not only discussed and analyzed in print but also dramatized. After *Faust, Cinderella* was rich Americans' favorite stage story: charitable amateurs, such as the King's Daughters of the Lenox Avenue Union Church, regularly "rendered the operetta *Cinderella* in costume," and Broadway productions were numerous. During April and May of 1894, for example, Abbey's Theatre, Broadway and Thirty-eighth Street, was offering Mr. Oscar Barrett's Fairy Extravaganza, *Cinderella*, direct from Mr. Henry Irving's Lyceum Theatre, and "social leaders" were "most frequently seen at the performances, their love of childhood's fairy tales having been revived to a great degree." *Cinderella*, after all, was more than a story about such culturally potent matters as a ball, a sudden rise from poverty to wealth, and scorn for unbeautiful women; it was a story about a foot.[27]

The foot, however, also occupied secret places in America, such as the literature of flagellation, the most popular perversion of the late nineteenth century. An anonymous 1907 enthusiast declared that "it is in the United States that the birching discipline is best known and most popular, being carried out with artistic, poetical sentiment until it becomes the inseparable, supreme refinement of love," and such surviving texts from the literature as *The Beautiful Flagellants of New York* and *Lashed into Lust* feature considerable foot and toe action. Bouguereau in 1870 had attempted to play directly to interest in flagellation when he painted a *Flagellation of Christ*, but the painting was heavily criticized for displaying too much "human anguish." From that point onward, Bouguereau confined himself to more subtle signals, especially bare feet in the foreground of a painting, signals that those in the know could pick up and those not in the know could miss entirely.[28]

With the publication of George Du Maurier's *Trilby* in 1894, feet emerged as an overwhelming public enthusiasm, legitimated by connection with France and art and love. The artist's model Trilby O'Ferrall, a "very tall and fully developed young female" who often wears men's clothing and who always rolls her own, first brings herself to the notice of Little Billee when she displays to him "her astonishingly beautiful feet, such as one only sees in pictures and statues—a true inspiration of shape and color, all made up of delicate lengths and subtly-modulated curves and noble straightnesses and happy little dimpled arrangements in innocent young pink and white." Little Billee is "quite bewildered to find that a real, bare, live human foot could be such a charming object to look at"; he pants as heavily as does Du Maurier's prose; he is in love.[29]

Foot fervor hit big-city culture very hard. The *New York Herald* was fascinated with feet and voluble on the subject of how to read "character" from them. In September of 1894, a *Herald* reporter sat all day in the lobby of the Fifth Avenue Hotel studying "Some Great Men's Feet" and announced some "interesting glimpses at the dispositions" of those whose feet he had studied. In November, the same *Herald* reporter was sent to study "pedology" at the horse show, where he was "much struck with the dudes' foot wear," and in March of 1895 the *Herald* announced:

FEET AN INDEX TO CHARACTER.

In spite of distorting shoes their lines continue to tell the tale
of mental traits.

THE TRILBY TOES THE THING.

Different forms of heels, insteps, and toes and what it is they all signify.

Not only were feet revealing, but to reveal them was to arouse the male populace. In 1895, in a shoemaker's shop in New Jersey, John Dwyer, a rum salesman, saw the "twinkling delicate pedals" of Miss Lillian Kane "flitting in and out of her fluffy skirts like tiny mice" while she waited for her shoes to be half-soled. He became so aroused that his eyes bulged, his face reddened, and "with a leap he scaled the counter and embraced the girl." Miss Kane pierced him with her hat pin, the shoemaker ejected him from the shop, and he was charged with assault.[30]

Trilby played directly to end-of-the-century American fascination with bohemian life in Paris; the stage version, which opened in March of 1895, was so popular that during the next year twenty-four simultaneous touring productions spread the news about feet across the United States. For a time, feet were regularly referred to as Trilbys in the pages of the newspapers, and Trilby contests (female contestants only) became a familiar cultural feature that drew large male audiences eager to officially assess foot beauty. By 1898, however, Trilby had become culturally confused with Cinderella, and Trilby contests were transformed into tiniest-foot contests, whereas Trilby herself had been equipped with a foot "neither large nor small." The medical eye also focused on the foot: *Punch* caricaturist Harry Furniss visited New York in 1895 and wrote:

> If I were asked what impressed me most about New York, I should not say Brooklyn Bridge, or Wall Street, but the number of chiropodists' advertisements! They confront you at every turn; these huge gilded models of feet outside the chiropodists' establishments, many painted realistically and many adorned with bunions. Now why is this? Do they squeeze their feet into boots too small for them, or are their pedal coverings badly made, or does the secret lie in the rough pavements of their thoroughfares? New York ought to be called New Trilby.

After George Du Maurier died in 1896, *Trilby*'s popularity plummeted, multiple copies sat untouched on lending-library shelves, and the text moved underground to become the stuff of complicated dirty jokes at stag parties. In 1900, the Trilby foot disappeared; "your Trilby" was no longer your foot but your hat. Although no hat figures in Du Maurier's *Trilby*, the original London production featured a hat and thus displaced focus from the pedal extremities to the thinking end of the body. The hat survived throughout the twentieth century, as did Svengali, the mesmeric controller of the novel; the foot went back into hiding.[31]

Salon art bared far more than toes, but Paris Salon paintings of female nudes were comparatively slow to arrive in America. When the female nude crossed cultures from France to America, she moved from a culture wherein she was commonplace to one wherein she was not, into an American atmosphere compounded of sexual reticence, ignorance of female anatomy and physiology, limited acceptance of "the classical," obsessive attention to women's looks, confusion of ideal bodies with real bodies, fears about innocence and corruption, and a firm class-crossing belief that visual representations could be morally ruinous to viewers. The standard art-history distinction between the naked and the nude did not cross the Atlantic along with the pictures, and in America that distinction did not necessarily serve as a framework for viewing. Nudes on canvas—not in marble, considered chaste—were thought especially unsuitable for women to see, even though women were considerably more familiar than were men with what women's bodies looked like. Men's effort to exclude women from the sight of a female nude speaks volumes about what late-nineteenth-century men were up to when they gazed at nudes. When a private collection containing even one nude went up for sale, the preview was advertised as open "to gentlemen only." Urban bars—all-male venues in any case—displaying major art collections did a reverse accommodation and advertised "Ladies' Day." At 11 A.M. on a Thursday, before the drinkers took up their positions, a woman might arrive at the Hoffman House, pick up a catalogue titled *Gems of the Hoffman House Collection*, enter a barroom emptied of all the usual males, and view the pictures. Such "special opportunities" not only reinforced the exclusionary nature of the "gallery" but also suggested great social fear and confusion: in a men's bar, a man could look at representations of nude women but could not look at a clothed woman who was looking at those same representations.[32]

The first gallery that Earl Shinn visited for the three-volume *Art Treasures of America* was the Corcoran in Washington, D.C., and it was there, at the start of his journey through American Salon art collections, that he deployed the oldest commonplace on record about French Salon nudes: that these pictures used scenes and figures from classical mythology as a cover for the depiction of nude bodies and as a shield against criticism. Examining Louis Priou's *A Family of Satyrs*, a painting whose foreground is dominated by a female nude, Shinn explained it as "the work of a painter who has made the revival of the mythical sylvan deities a specialty. By taking up this class of subjects he secures the privilege of studying the nude without blame." That much-repeated assertion fails utterly to explain the centrality of the female nude in Salon art. The female nude appeared, after all, not only in "classical" pictures but in French Orientalist harem scenes, bathhouse pictures, and fantastic representations claimed to be Moorish slave auctions. In America it was a social fact that Salon nudes were lodged entirely in the territory of men. Men commissioned the pictures, dealt in them, draped them in bowers of pink fabric, threw spotlights on them, displayed them to each other, passed them from hand to hand, blew cigar smoke at them, imbibed gin slings and corpse-revivers in their presence, and reproduced them on cigar-box lids and tin signs advertising whiskey.[33]

In print, the subject of the nude was marked by either silence or, occasionally, denial of presence. In 1894, the *New York Herald* announced a portrait show of society belles under the headline NO NUDES HERE. The headline addressed a male reader who might well assume a woman rendered on canvas to be nude unless the *Herald* disabused him of that idea. Then again perhaps this particular show was notable chiefly because the female subjects kept their clothes on; therefore a man looking to see some female nudes need not bother with this show. Sometimes reticence took a different turn when art critics simultaneously articulated and failed to notice the subject of a painting. In 1894, in a review of the Salon of the Champs Elysées, the *Herald* paid special notice to "a most attractive exhibit"—a rape scene and a torture scene—from the brush of Evariste Luminais: "His *Norman Pirates in the Ninth Century* represents two Norsemen abducting a woman on the coast of France. They are carrying her toward one of their vessels and are knee deep in water. She is making fruitless efforts to escape. *The End of Queen Brunchaut* illustrates the legend of a queen who, in the seventh century, was condemned to be tied to a horse's tail as a punishment for her numerous crimes. The coloring is marvellous." The French novelist Paul Bourget, noting in America the "entire absence" from American publications of both nude drawings and stories of "marital misadventures," concluded that all matters sexual existed "in such a shadow of secrecy that in America they escape even satire."[34]

In this volatile American mix of fascination and silence, display and concealment, a man who could get his hands on the right French nude, and who had the "gallery" to display it to other men, could go a long way. In 1871, Edward Stiles Stokes murdered James Fisk, Jr., the flamboyant Prince of Erie and gargantuan swindler, on a public staircase in the Grand Central Hotel in view of two witnesses. Edward Stokes was equipped in life with not much more than good looks, rich relatives, and a profound understanding of money and consumer icons. A reporter who visited Stokes in the Tombs, where he awaited trial, found him wearing lavender trousers, a silk velvet dressing jacket trimmed in pink quilting, slippers embroidered with gold lace, and a diamond solitaire. He had a carpet on the floor, he had his meals brought in, and when the opportunity arose he asked the warden to light his cigar for him. Edward Stokes knew that no man backed by money had ever been capitally punished in the United States, and he also knew the iconic power of diamonds and velvet. He escaped capital punishment. Five years later, when he was discharged from the state prison at Auburn, New York, he deployed the icons again: he emerged from the big house wearing a diamond stud and a black overcoat faced with black silk velvet.

EDWARD S. STOKES

When Stokes reappeared in Manhattan, however, he learned that despite such displays on his part, no man would shake his hand. He was an outcast. Had the year been 1893, by which time public figures were seeking to have personal histories of horse theft, embezzlement, and murder rewritten as "schoolboy episodes," things might have been different for Stokes.[35] The year, however, was 1876, and so Stokes trained himself to rise and nod, but never to court rejection with an extended hand. For the rest of his life he held himself aloof and kept his back to the wall. He would not reestablish himself in Manhattan until 1882, when, backed by Cassius Read at the Hoffman House and by a partnership with John Mackay of the Comstock Lode bonanza, he bought at auction a huge and glowing French Salon painting stocked with five life-size nude figures: Adolphe-William Bouguereau's *Nymphs and Satyr* (1873) for $10,010.

Thirteen years earlier, in 1879, Earl Shinn had visited the private collection that held *Nymphs and Satyr*. John Wolfe, then the owner, was "not afraid of big pictures," wrote Shinn. The walls of Wolfe's house were stacked Salon-style with ninety-six "ample" paintings hanging "well crowded up, from the ceiling down—great draperies of canvas, on which the eye makes out contours and attitudes of life-size figures; not one or two sheets together, but a whole wall of them, and then, at right angles, another wall similarly paneled. The huge gilded frames rise to the cornice like pilasters, rubbing their fretted edges or parting with their external moldings to bury themselves in reserved space." In general, Earl Shinn disliked Bouguereau's work. He thought that John Jacob Astor's example, *La Tricoteuse*, exemplified "the inherent character of Bouguereau's style, as if the canvas had been scraped all over with a piece of glass, and then waxed like a bit of Queen Anne furniture." "Waxen" was the term he applied most frequently to the Bouguereaus he saw on his tour of the galleries of the rich, and generally he did not even pause to describe the paintings: "A word and a line, a phrase and a hasty diagram, are enough to elucidate the specimen," he wrote on seeing A. T. Stewart's example. "Bouguereau's peculiar talent being given, and the scale being mentioned as life-size, 'you see from here,' as the French say, just how he would treat it." Across the three decades of Bouguereau's overwhelming popularity in the United States, influential American art-writers joined Shinn in a community of boredom with Bouguereau: Clarence Cook called him "an industrious plodder, always pursuing the same subjects and painting them in the same way," C. H. Stranahan found

Bouguereau's pictures "sincerely destitute of feeling" and classed him as no more than a successful decorator, and W. C. Brownell simply took a pass when faced with a Bouguereau: "Life is short," he wrote, "and things of more significant import demand attention."[36]

When Earl Shinn saw *Nymphs and Satyr*, however, he did a turn-about. It was, he wrote, "so unexpectedly fine an exception that one is tempted to drop the lance which habitual prejudice puts into the hands of anybody who is taken to admire a Bouguereau. Four life-size women of the woods have caught a goat-faced satyr at a disadvantage, and are pulling him into the water by the arms, the ears, and the horns." Astonished, Shinn praised the picture's "precipitate action," its "combination of life and spirit," and the "polished back" of the nymph at the lower right—later characterized by enchanted viewers as the most gorgeously lit set of buttocks on canvas. Shinn's sole quarrel with the painting was that the women looked like "ladies, not Mænads or Bacchants" who had "accidentally" got undressed while "reading the French fashion-newspapers." Shinn did not venture a guess at what the satyr had been reading.[37]

In 1882, John Wolfe put the collection containing *Nymphs and Satyr* up at auction, and the *New York Herald's* art reporter who attended the presale "private view for gentlemen only" was simply knocked over by the picture.

> The triumph in the collection is for Bouguereau, who in cen-turies to come will be as much esteemed as many of the old masters are today. Call him cold; blame him for lack of strength, for the extreme suavity and waxiness if you will, his drawing, modeling, composition and selection of charming types will make him renowned, when possibly Henner is almost forgotten and Manet shall have become but a name. The *Nymphs and Satyr* is one of his best works, executed in his prime, notwithstanding its suave licked appearance and some-what bloodless, though otherwise very real, quality of flesh.

After lingering over the "luscious band" of nymphs in the picture, the *Herald's* man was suddenly moved to announce, in the smoke-wreathed men-only context of the preview, that "every woman should be proud of this work, the apotheosis of the female form divine. It would have been purer still"—signaling that purity was the unannounced subject here—"if the blue and red fillets had been omitted from the heads of two of the nymphs."[38]

Edward Stokes transported his prize picture to the Hoffman House on Madison Square at Twenty-fifth and Broadway. In 1882, the Hoffman House, a hotel regularly listed in 1880s guidebooks as "strictly first-class," was booming. An "Italian Renaissance" annex was under construction, the Democratic Party used the hotel as its unofficial headquarters, and in the splendid Hoffman House barroom, said to be seventy feet by fifty feet, politicians, businessmen, theater people, and "sporting men" gathered daily. Occupying opposite corners on Twenty-sixth Street were the Café Brunswick and Delmonico's, and nearby was the Fifth Avenue Hotel, headquarters for the Republican Party. All told, Madison Square was a center of men's action at that moment, and Stokes raised the pitch when he installed *Nymphs and Satyr* directly opposite the mahogany bar of the Hoffman House: the painting appeared twice, as itself and as reflected in the mirror over the bar. Stokes, whose sense of the icons was unerring, positioned above the picture a swagged red velvet canopy and before it a crystal chandelier; a low brass railing kept patrons at a respectful four-foot distance from the painting, and a brass spittoon directly in front of the railing signaled that viewers were in the domain of men. A wall plaque announced the price Stokes had paid for the Bouguereau.

At that moment French Salon art made its real public debut in America, and made it with a never-rivaled and astonishing example of what Salon art offered: a big canvas, soft lighting, pleasant flesh, a suspended story line, and a hint of the "classical." Gradually, Stokes and his partner, Cassius Read, filled the barroom with ceramic "Nubian" statues, taxidermized trophy heads, plants claimed to be rare, and more paintings, including Luis Falero's *The Vision of Faust*, in which an orgiastic tangle of two dozen naked bodies streams past the sleeping Faust, supervised by Mephistopheles. The finished effect Stokes and Read announced to be the Hoffman House Art Gallery.

Adolphe-William Bouguereau's *Nymphs and Satyr* is a frozen whirl of motion, a tight spiral around the satyr whom the nymphs are attempting to pull into a stream. At the lower right of the spiral, a golden light illumines the buttocks of one nymph, then glances off the forearm of the nymph on the left, and finally touches, at the top of the spiral, the left hand of a third as she beckons to more nymphs in the background. All four nymphs are perhaps one nymph, one model, one placid, pleasant, squeezable girl with breasts of a most perfect symmetry. A leafy branch here and a satyric arm there prevent total nudity. Although the girls have fastened on the satyr's arms, hair, and horn, they do not look athletic or muscular; they do not look like swimmers. The action at the moment is a mock assault, a mere suggestion of what might happen if anyone chose to get serious. Although the satyr may feel more threatened by the water than by the nymphs, the male viewer lounging against the bar, foot up on the rail, was unthreatened. If one is a male unmarked by cloven hooves and woolly coat, the picture signals a wonderful pleasure: four pretty girls, all smiling, all naked, tugging at just one fellow. There is no danger; everyone will survive to smile and gambol another day. It is a fantasy of multiple sex partners, of pretty girls ready to have fun, of gravity-defying breasts. Pleasure is far more likely than harm.

"That French 'freedom,' " wrote Clarence Cook, "which offends so many people here in America is never betrayed in Bouguereau's pictures; they are as pure, as passionless and as cold as an anchorite could desire." Perhaps so, but men in the Hoffman House both yearned over and laughed at the painting. Like Louis Priou, everyone in the Hoffman House bar "studied the nude without blame," and studied some other subjects, too, for *Nymphs and Satyr* is not a simple "nude." Satyrs are half man, half goat, in love with music, revelry, and lechery; nymphs are, among other things, the objects of satyrs' lust. In those contexts, the Bouguereau spoke directly to the time during which it was on display at the Hoffman House. In a world that validated the way of life of the man-about-town, *Nymphs and Satyr* was a pretty reversal of the actualities of sexual predation; it altered the more usual direction of the assault. *Nymph* in New York City in the 1880s and 1890s was a standard term for a prostitute, *nymphs du pave*, as the lawyer Abe Hummel liked to call them. The *Nymph* label was tacked onto various depictions of nude women, and the nymph-and-satyr combination was especially favored among certain art collectors. After man-about-town Stanford White was murdered in public by Harry Thaw in 1906, among the artwork retrieved from the city hideaways where White conducted his sex life were numerous nymph-and-satyr combinations. Albert Koller's *Nymph and Satyr* offered a "figure of a woman reclining at full length on sumptuous draperies . . . turning her head toward a dark-skinned satyr who crouches just beyond and near her," and an Early French School *Nymphs and Satyr* depicted "an agile nymph, partly draped in a flowing blue mantle, who has succeeded in escaping from the clutches of a satyr, who has fallen prone, holding in his left hand a tress of hair which he has torn from the head of his intended victim." Stanford White's male friends went to the sale and bought the pictures.[39]

Tourists and travel writers, especially those from abroad whose knowledge of Stokes's past was dim or confused, discovered the Hoffman House and, whether or not they liked what they saw, helped to make it internationally famous by arguing about it. In 1882, T. S. Hudson admired the "quiet elegance" of the hotel, and described *Nymphs and Satyr* as "a handsome oil painting, the subject of which is so 'classic' that I dare not describe it here." In 1884, however, Lepel Griffin thought the hotel decor overdone and dropped the Bouguereau into the category of things that "Americans are satisfied with because they are large; and if not large, they must have cost a great deal of money." William Hardman, in the same year, was simply astonished by the Hoffman House bar, "the most splendid drinking saloon in the world." Hardman and his tour group, which contained "ladies," at first hesitated to enter the bar and "tried to obtain a view of the brilliantly lighted interior through the large uncurtained windows, but the manager [possibly Edward Stokes, since he held that title at the Hoffman House] seeing us, and that we were a party of foreigners, came out, and begged us all to come in, assuring us, as was quite true, that although the saloon was for men alone, still there was nothing to cause the least offense to ladies. If we had not accepted his invitation we should have missed a very great treat."[40]

William Hardman judged the Hoffman House artwork, all mounted on crimson velvet and each with its own electric light to show it off, "almost priceless," disregarding the fact that the prices were displayed next to the artgoods. *Nymphs and Satyr* he found to be a "grand painting," and he was further impressed by "lovely statuary in marble and bronze, a large and unique piece of Gobelin tapestry made for Napoleon III, a very choice selection of articles of *vertu* and rare plants," and "a genuine Correggio of great value . . . an undoubted original . . . the subject is Narcissus." Every claim that Hardman heard becomes shaky when the "genuine Correggio" appears on his list. Correggio never painted a Narcissus, but Edward Stokes always believed that a big assertion could capture people's attention and per- haps even their respect. The art and money combination was so potent a validation that Stokes took to claiming he had been offered the equivalent of half a million for the Bouguereau. That claim caused customers' jaws to drop. That Bouguereau was valuable, and Stokes, as long as he stayed within his gallery, once again also had value.[41]

Americans were hungry to consume visual images, and the pleasing consumer experience of the personal art gallery spread across the country, its sites and scale shifting as it moved. Galleries appeared not only in urban saloons but also in hotels, in private big-city clubs, in brothels, in restaurants, in theaters, in gambling houses, in doctors' offices, and in the houses of the very rich, the merely rich, and the aspiring. In cramped quarters, persons with only a corner to spare for art tipped the gallery wall, figuratively speaking, and created a flat tabletop "art unit." All galleries, however, adhered to identical princi- ples of crowding and variety: a private collection should fill the space allotted, and the pictures should stand or hang edge to edge. A gallery required a great many pictures and objects: even if no single object in the collection could stand on its own, and even if all the objects were individually dubious, a space crammed with them nonetheless quali- fied as a collection and thus averted critique. No two pictures or objects should be identical; if there were two reproductions of the same work, they should be of different scale or material. Nothing, even if once designed for use, should appear in usable condition, and there should be no center: the eye must not be drawn toward a central image but kept moving from one image to the next.[42]

If the urban private galleries of the rich were spread under a skylit dimness, other types of galleries occupied a corresponding social twilight. During the 1894 Lexow Committee investigation of corruption in the New York City Police Department, John W. Goff questioned Captain Max Schmittberger of the Nineteenth Precinct about "two houses" once kept by Emil Patel at 102 and 104 West Twenty-seventh Street.

Q. That was a notorious place, was it not?
A. Yes, sir.
Q. Did you ever hear that house called the captain's house?
A. No.
Q. Did you ever hear of that house having an art gallery for exhibition where men about town and strangers visiting New York were taken to look at those pictures?
A. No, I don't remember that; I thought I knew all the art galleries in the Tenderloin; I never heard of that; it is possible.
Q. We will call it an album of French pictures?
A. Oh, yes; an album, there were French pictures there, yes, sir.
Q. We simply misunderstood each other about the term art gallery.[43]

The lawyer and first-nighter Abe Hummel papered the walls of his inner office with "affectionately inscribed portraits of stage beauties," and many men-about-town kept private scrapbook galleries of nude photos, pictures of showgirls, and "art work" of women. In 1891, the *National Police Gazette* put on sale a gallery of two hundred photographs, a dime apiece, "finished in sepia and suitable for framing," of prizefighters, billiard champions, and actresses available in three poses: "in costume, in tights, or bust showing." A gallery could arrive annually in the mail: *La Nue au Salon* offered its subscribers reproductions of all Salon nudes for a given year. The 1902 peepshow book *Famous Paintings* offered another preassembled gallery: 341 "pictures reproduced at an enormous expense from great paintings of the principal art galleries of the world." *Famous Paintings* invoked the "well known collections of George Vanderbilt" as context for an exhibition of Salon-style Psyches, Vestal Virgins, and King's Favorites that resulted in a display total of 364 bare breasts, 67 cupids, 45 wisps of gauze, a great many raised arms, and an erotic favorite, much waist-length loosely flowing "tressled" hair. Least visible of all the galleries was that inked onto the human epidermis: during the tattooing craze of 1893, according to the *Herald*, "any number of men in this city have been beautifully decorated and are simply walking picture galleries."[44]

The magic of all things French began to fade at the end of the century, but the American vogue for Salon art did not so much fade as collapse, in the process sending up a small puffy cloud of mystery and embarrassment. In *The Strange Life of Objects*, Maurice Rheims, remarking on the "final disappearance of conventional values in art," writes:

> After the prices of the old-school paintings had collapsed, these pictures virtually vanished from the market, to the extent that it became almost impossible to distinguish what might have been important in that period; which in turn only helped to discredit it still more.[45]

The French Ball collapsed under the weight of its own notoriety. Burgeoning antivice societies seeking to clean the nineteenth century's slate made the French Ball such a special focus of preventive effort that by 1897 there were more censorious observers at the ball than there were uninhibited revelers. In that year, Cholly Knickerbocker of the *Journal* and dozens of other New York City "chappies" who had looked forward to the French Ball were miserably disappointed:

> Every chappie that I saw yesterday declared that the ball was dull. It was worse. It was a bore. For fifteen years I have seen

the annual *bal masque* of the *Cercle Française de l'Harmonie*,
and in that time it has shrunk from a carnival to a carouse.
Monday night scores on scores of the toughest-looking citizens stood about the promenade with hats and overcoats on
and with cigars tip-tilted in the corners of their bulldog
mouths. They made no pretence at costume or evening dress
and were evidently too poor or too mean to buy hat checks.
They may have been policemen out of uniform, but whatever
they were they cast a heavy gloom. That is the sort of thing
the chappies won't have. Some of them will stay till the crack
of doom in the vain hope of seeing the tough woman, but
when the tough man is *en évidence*, they want to go home.[46]

Cholly Knickerbocker "fancied that in this condition of things I saw
the doom of the French ball." He was right. A number of causes
worked to finally shut down the French Ball in 1901. The mere word
French had lost its power as a cover, the antivice crowd was increasingly successful at clearing the way for the arrival of the puritanical
twentieth century—and big-city men were arranging to meet "tough
women" in private and protected venues, there to experience the
pleasures that the French Ball had made all too public.

NOTES

1. George Inness, who lived and worked in Europe, was the sole exception
to rich buyers' exclusion of American artists.

2. *Chicago Tribune,* 11 June 1893; Fraser, 243. In *A World of Goods,* Mary
Douglas defines modes of exclusion that interestingly illuminate the art-collecting world of the late nineteenth century: "One way to maintain a social
boundary is to demand an enormous fee for admission. Another is to set the
normal rate for settling of internal transactions so high that only the very
rich can afford to join the game." The third, and most telling mode is the
"refusal to transact . . . a common, if not worldwide, strategy of exclusion"
(140).

3. *New York Times,* 9 January 1897; *New York Herald,* 10 February 1895;
Burne-Jones, 215; *New York Herald,* 17 December 1893, 19 February 1893.

4. "Good Art as an Advertisement," 111–115; *New York Times,* 11 January 1897.

5. *New York Herald,* 13 January 1894.

6. *New York Herald,* 17 February 1907, 27 August 1899.

7. *New York Herald,* 16 May 1894.

8. Van Dyke, 376–377.

9. Sheldon, I, 118–119; DeLisle, 328–334.

10. *New York Times,* 16 January 1897; *New York Herald,* 1 March 1893, 31 January 1894.

11. Tod, 55–66. On March 9, 1902, Thomas Kirby, the auctioneer, told the *New York Tribune:* "Every sale of good pictures increases the general interest in art. . . . I believe that the dispersal of art collections does more in this way than the great museums. The latter stimulate art interest among those who cannot afford to own paintings, but the rich man who may become an art patron does not take the time and trouble to study the public collections. He is interested in the paintings which hang in his own gallery if he has one, and those in the galleries of his friends. The spirit of rivalry enters into the matter and the purchase of first class examples of art no longer becomes a question of cost. There is a constantly increasing demand for fine paintings, owing to the fact that our rich men are building big houses which they desire to embellish."

12. Whibley, 11; *New York Herald,* 23 January 1907; Bourget, 146; Soissons, 79; Naylor, 98.

13. *New York Sun,* 25 January 1920; Howe, II, 124.

14. Morgan, *New Muses,* 71; Day, I, 20; *New York Herald,* 10 December 1893.

15. Lind, 152; Gilfoyle, 176.

16. The Lexow Committee investigated police corruption in New York City; it was named after Clarence Lexow, New York state senator from the Sixteenth District.

17. Lexow Committee, III, 3118–3126; IV, 4579; Gilfoyle, 282.

18. *New York Herald,* 12 February 1893, 21 February 1893; *New York Times,* 15 February 1893.

19. *New York Herald,* 11 February 1894.

20. *New York Herald,* 12 February 1895, 17 February 1895.

21. Seventy-nine American artists appear in the index to Shinn's *Art Treasures of America.*

22. Rheims, 198–199; Constable, 73; *New York Herald,* 1 April 1893, 11 March 1893. Fakes and forgeries of "Old Masters" and of Corot, mixed into the Salon art scene, further muddied the art-collecting situation. W. G. Constable wrote in 1902 that "it came to be said that Corot painted two thousand

pictures, of which three thousand were in the United States," a great many of them from "a flourishing Corot manufactory in New York," but individual buyers did not seem to feel personally devalued when their picture was devalued, and they did not necessarily stop displaying and enjoying their Corot just because they had learned it to be faked. In "Campbell Corot," a story in Frank Jewett Mather's *The Collectors,* Campbell Corot assesses his own lifetime output of fake Corots: "The best things I do, or rather did, young feller, are jest a little poorer than his worst. Mine are plenty good for everyday purposes. Nobody can paint like his best" (22–23). When an entire collection of certified fakes, such as the 205 paintings belonging to Dr. James A. Leaming, went up for auction, the effect was both touching and hilarious: the late Dr. Leaming was characterized as a "sincere lover of art" even as his "Raphael" went out the door for $77.50.

23. Constable, 69–71. In *The World of Goods,* Mary Douglas discusses luxuries as consumer goods, and notes, on the one hand, that demand for luxuries must be diversified, with each item sending its own signal; on the other hand, those luxuries tend to become standardized. At that point, at "the hot center" of a competitive system of consumption, "some crucial form of social control is being exerted" (144–145).

24. Cook, *Art and Artists of Our Time,* I, 88.

25. *New York Herald,* 10 April 1894. Evelyn Nesbit united in her person several top sexual attractors of the time. Besides her proclivity for the peasant blouse, she was said to have "the most perfectly modeled foot since Venus" (O'Connor, 187), she had a terrifically thick head of hair, and she married Harry Kendall Thaw, a dedicated flagellator who beat the skin right off her back.

26. Cook, *Art and Artists of Our Time,* I, 85–88. Menard quoted in Shinn, I, 80–81; Beckwith, 262; Shinn, II, 67; I, 122; Davis, 110.

27. *New York Herald,* 16 April 1893, 13 May 1894.

28. Hay, 218–221.

29. Du Maurier, 12–15.

30. *New York Herald,* 30 September 1894, 25 November 1894, 10 March 1895, 15 December 1895. Men were in thrall not only to women's feet but also to their shoes. In 1893, after the grand-scale embezzler Francis Weeks ran off to Costa Rica, officials in search of evidence broke into his desk, apparently because it was so heavy that they "believed it to contain money." In the desk they found twenty pairs of women's shoes plus uncounted numbers of single shoes. Weeks, judged the *Herald* on 16 October 1893, "had given himself up to the shoe language as a poet yields himself to the witchery of dazzling eyes." When that poetic analysis collapsed, the *Herald* turned to the language of

business crime: "The discoverers of this treasure trove were at a loss to know what had prompted this strange fancy . . . and not a little puzzled to know how Weeks had found opportunity to embezzle the shoes of fair women to such an alarming extent."

31. Furniss, II, 57–58.

32. In *The Nude,* Kenneth Clark offers the standard distinction: "To be naked is to be deprived of our clothes. The word nude, on the other hand, carries, in educated usage, no uncomfortable overtone."

33. Shinn, I, 9. Jennifer L. Shaw does a full treatment of this subject in "The Figure of Venus: Rhetoric of the Ideal and the Salon of 1863," *Art History* 14, no. 4 (December 1991), 540–570.

34. *New York Herald,* 4 November 1894; Bourget, 343–344.

35. *New York Tribune,* 18 April 1893.

36. Shinn, I, 43–44; II, 11; *Masters in Art,* 34. Earl Shinn thought that *Nymphs and Satyr* was 5'×10'. Later viewers believed it to be 8'×10' and even 8'×12'. Currently it hangs in the Sterling and Francine Clark Art Institute in Williamstown, Massachusetts, where its measurements are 8.5'×5.8'. In 1906, Bouguereau's works were almost entirely in private collections in the United States; they were difficult to trace and were constantly changing hands. Only six were listed as being in American museums in 1906. By 1999, the compilers of the current Bouguereau catalogue raisonné had documented 791 surviving Bouguereaus in the world.

37. Shinn, I, 53–54.

38. *New York Herald,* 12 March 1882.

39. Cook, *Art and Artists of Our Time,* I, 85; *Howe and Hummel,* 51; *Old and Modern Paintings, Water Colors, and Drawings Belonging to the Estate of the Late Stanford White.*

40. Hudson, 38–39; Griffin, 17–18.

41. Hardman, 31–32; Mulhall, 129. Mulhall says that at the Hoffman House's "magnificent art gallery . . . one day in the week was 'Ladies' Day,' and any lady calling at the hotel was met by a boy with a printed catalogue describing the house and its works of art. Ladies were escorted through the bar." The concept gained currency, and by 1891 there were numbers of such bargalleries in the city. In *A Week in New York,* Ernest Ingersoll noted that "Stewart's, at 8 Warren Street, has some paintings of great merit, which are open to the inspection of ladies from 9 to 11 o'clock in the morning" (87). After the Bouguereau vanished from sight, its price continued to rise: in 1918, Arthur Maurice called it "the twenty-five-thousand-dollar [equivalent of half a million] Bouguereau" (330).

42. The flat art unit must have been a challenging element in home décor.

When, in an effort to understand how such a standard feature of nineteenth-century life worked, I decided to build such a unit, I found that the space I had dedicated to it, although only twenty-nine inches by eleven inches, was impossible to fill. No matter how many plaster Venuses or pictures on little gilded easels or "ceramic attractions" I put into the space, it continued to require more.

43. Lexow Committee, V, 5333–5334.

44. Rovere, 23, 41; *Famous Pictures,* throughout; *New York Herald,* 18 April 1893.

45. Rheims, 198–199. Public disappearance of Salon-style art was pushed along whenever Anthony Comstock, secretary of the Society for the Suppression of Vice—one of the many preventive societies that seized legal authority for themselves in the 1890s—went on another of his rampages against Salon nudes. On the Brooklyn Elevated Railroad in 1895, Comstock spotted an advertising placard, twenty-four inches by fourteen, that reproduced Luis Falero's watercolor *Twin Stars* (1881), a very well known item in the Catharine Lorillard Wolfe collection at the Metropolitan Museum of Art, and still owned by the Met. The original "represented two nude figures, women, floating through a midnight sky, their index fingers pointing upward toward two brilliant stars directly above them. Comstock characterized the placards as "a menace to the public morals" and "suggested in an authoritative tone that everybody and everything would be better served by their instant removal from the cars." The cards were not removed but patched with large squares of black paper, and thereby became "the main subject of conversation and amusement for all. Mr. Comstock is not spared the ridicule that American crowds indulge in." An official explained to the *Herald* that the railroad wanted "no trouble with Mr. Comstock. We did not see anything wrong with the pictures ourselves, but if Mr. Comstock did we were willing to take them down and save unpleasantness that might otherwise ensue." As usual, Comstock was ridiculed by press and public, but also as usual he had his way (*New York Herald,* 28 December 1895). In *Intimate Matters,* John D'Emilio and Estelle Freedman demonstrate that Comstock, by suppressing all information on sexuality and reproduction, drove sexuality into so private a sphere that sexual behavior became far more risky for women than it had been. Comstock, in effect, made the early decades of the twentieth century considerably more repressive than the nineteenth had been (160ff.).

46. *New York Journal,* 20 January 1897.

CHAPTER 3

Bargains in Bric-à-brac

Cosmopolitanism is more dearly bought than we at first imagine.
—Emily Faithfull, *Three Visits to America*, 1884

IN MARCH OF 1895, Blanche de Berzsonyi, by her account an Austrian baroness, disembarked unnoticed in New York City. She settled into a suite at the Waldorf wherein she considered how she might sell to rich Americans the contents of her luggage: a heap of "Chinese, Japanese, and Oriental curios and articles of bric-à-brac."[1] The Baroness also had a heap of stories about both her own identity and her connections to the American rich. She told the *New York Times* that Alva Vanderbilt was an acquaintance of hers and had asked to "inspect" the artgoods; she told the *New York Herald* that "friends" had suggested Alva Vanderbilt as a likely purchaser. Whatever the case, the Baroness dropped by the Savoy at Fifth Avenue and Fifty-ninth Street, where Mrs. Vanderbilt kept a suite for just such shopping occasions, and left there "some of the most suitable of the bric-à-brac" for Mrs. Vanderbilt's inspection.

Notable among the Baroness's bric-à-brac was a Tantalus cup, a "philosophical toy" about a foot high, "containing the figure of a man, in whose body is concealed a siphon, which prevents the fluid from passing out when a person attempts to drink from it." The cup was supposed to exhibit the fate of Tantalus, a Lydian king who was admitted to the banquet of the gods and then, charged with having betrayed their secrets, was condemned by them to suffer constant hunger and thirst. According to the Baroness, the cup was a rare possession of "exquisite workmanship and oddity and, moreover, owing to the fact that there are but few of the kind and quality in existence, was reasonably worth the sum of $500."

Blanche de Berzsonyi's title could carry her to the point of having her goods inspected, but not necessarily to the point of sale: when Alva Vanderbilt spotted nothing desirable among the bric-à-brac, she notified the Baroness to drop by the Savoy and reclaim her goods. The Tantalus cup, however, was no longer in the heap, and when the Baroness called attention to its absence, Mrs. Vanderbilt "said she had no idea what had become of it. She thought perhaps it was lost or mislaid in the hotel, where a large number of valuables had been sent, that it had been left with her for inspection and that she assumed no responsibility for its care." Convinced that she had been robbed of her finest curio, and urged on by her "friends," the Baroness slapped a lawsuit on Alva Vanderbilt and then retreated to Europe. Satisfaction she would not get. In the storm of "rare and exquisite" goods from all over the globe available for consumption in 1895 America, the Tantalus cup did not stand out as especially valuable: it was a multiple, it was neither jeweled nor made of gold, it had never been owned by Marie Antoinette, and it had no famous European artist's name attached to it. On the American consumer scene, it was a trick cup that not only prevented consumption but also taught a lesson about consuming. As such, the Tantalus cup seemed quite out of tune with the times. When the Baroness explained to reporters that she ordinarily kept the delicate cup "in a plush-lined cabinet which was seldom opened" except for special festive events, she demonstrated her incomprehension of the culture to which she had carried her goods. The cup had made a transatlantic journey but its festive aura and its special cabinet had stayed behind in Europe.[2]

The Tantalus cup was just one curio lost in the deluge of decorative objects that American shoppers "inspected" in their frenzy to create special "effects" in the domestic interior. The national urge to decorate crossed class levels and consumed both genders: the urge, as James Muirhead noted in 1898, was "not confined to the circles of the millionaires but crops out more or less at all the different levels." In the cities, there were plenty of professional decorators ready to create the look of the moment, but the dominant presumption held that anyone armed with advice literature and shopping energy could create the right look unaided by hired experts. Shopping was its core activity, and shopping went on in good times and bad. In hard times, shopping, in fact, lifted one's spirits, as the *Herald* announced in an 1893 seen-in-the-shops piece:

> If anyone has the blues over the hard times and trembles for the present and future prosperity of the United States let him make a tour of the great shops of the city on one of these bright and deliciously cool September days, and however deeply financial pessimism may have crept into his soul he will experience a sensation of refreshment and will become firmly convinced that all hope is not yet lost.[3]

The domestic interior into which big-city Americans toted their purchases had been under heavy decorative pressure for most of the century. Décor in America had gone through revivals called Greek, Renaissance, Colonial, Gothic, and Rococo; had felt the impress of styles called Louis XIV, Louis XV, Second Empire, Japanese, and Moorish; and had been subjected to reforms called aesthetic, Eastlake, and Arts and Crafts. There had been not only revivals of revivals but also continual revaluations of styles. The American explosion of interior decorating touched much more than domestic interiors: intensely styled décor efforts pervaded hotels, theaters, restaurants, churches, sanitaria, doctors' waiting rooms, stores, business offices, armories, and brothels. Only the scale of the effects differed.

During an especially difficult period in the late 1870s and early 1880s, the dominant idea in décor required that each room be done in a different period style. George Sheldon's *Artistic Houses* of 1883 is a monument to that decorative moment and to the house whose occupants moved from an Early English Renaissance hall through a Louis XVI parlor, a Louis XIII library, a Henry IV dining room, an Anglo-Japanese sitting room, and a Chinese reception room, finally to seek rest in an Alpine bedroom or to breathe fresh air in a Roman roof garden. Interiors decorated on the *Artistic Houses* model were not only full but fully unlivable; occasionally, in search of livable space, their owners abandoned the finished results and moved into a modest apartment around the corner. Furthermore, no sooner did *Artistic Houses* see publication than its day was past, its intimidating décor notions dismissed as "servile" and "slavish." Bad words, those, on the American thoughtscape. The period-room movement eventually found its place in American museums, where velvet ropes kept the rooms unpeopled.[4]

Across the 1880s and into the 1890s, the national decorating frenzy exhausted styles with increasing speed. One day's must-have look abruptly lost appeal and became yesterday's tiresome stuff. In 1893, the *New York Herald*'s décor arbiter sighed over how suddenly a "pleasing innovation" could become "monotonously common by repetition—the fate of everything decorative." Furthermore, in the 1890s, big-city Americans came to want portable decorative effects, objects soft to the touch, and arrangements freed of the old 1880s stiffness. They wanted less a finished décor than an ongoing project. Late-nineteenth-century America was not, however, a discard society: transition from one style to the next was often achieved by addition, by smothering the previous style under the accretions of the next one. Rooms thickened with global accumulations of goods as decorating Americans observed the unquestioned first principle of late-nineteenth-century décor: MORE.[5]

Chaotic scenes occurred amid—and because of—densely packed domestic interiors. In September 1893, Joseph Stone of 229 Third Avenue was calling on a friend at 4 East Nineteenth Street when he suddenly rose and "attacked a pile of music on the piano and sent it flying in all directions." He was "a powerfully built fellow, in the prime of life," and no one knew how to stop his rampage. His brother dashed to the East Twenty-second Street police station for help, but the sergeant in charge there "declined to interfere," claimed that the case was "out of his precinct," and suggested the West Thirtieth Street station. By the time Stone's brother found a policeman and returned to East Nineteenth Street, Joseph Stone was "in undisputed possession of the parlor. There he stood in the midst of a wild confusion of torn music, books, and bric-à-brac." Laughing heartily at the havoc he had wrought, Stone indicated that the room had caused him to feel "crazy."[6]

Searching for a looser and more fluid—yet no less cluttered—style, décor arbiters looked to the artist's studio, a space that had fascinated Americans since the early 1880s. Studio style clustered, in just one space, unlimited goods plucked from any period and any culture, regardless of original use or purpose. Furthermore, it erased room divisions and concealed functions behind screens and curtains; consequently it could be imitated anywhere, whether in a house, an apartment, or a single boardinghouse room. Because studio style relied on portable *objets d'art*, pillows, hangings, and lamps, it could be replicated without furniture, and its flimsy plywood underpinnings could be concealed under heaps of decorative fabrics and Oriental rugs. By using every surface, including the floor, as display space, studio style extended the pleasures of continual rearrangement. Beyond all else, the style carried a dark aura of the *artistic,* the mysterious—the safely mysterious, it was believed—and the sensual.

In transfer to the living quarters of ordinary human beings, studio style jettisoned the artist and most of the artwork; it was itself the picture, and as such seemed to offer both unlimited visual stimulation and access to the sense of touch in a culture terribly worried about touch. Its promises were many and heady for anyone who was not an artist: unlimited shopping, happy ransacking of other cultures, and achieving the desirable décor of travel even if one was personally untraveled. Clarence Cook, the décor arbiter, suggested that studio style could confer on any interior "some of that freedom from conventionalities and old-time preciseness that is at least shadowed forth in the best of our artists' studios." Furthermore, the style emanated a nicely controllable risk. Initially no discussion of it suggested that anything in such an object-world might escape an individual's control, that in goods extracted from so many cultures there might still be lurking an unexpected power, that something might leap from a shadowy corner of the studio to overwhelm its owner.[7]

Unlike late-twentieth-century Americans, in whose bare living spaces the ravages of technology reduced visual stimulation to the contents of a single screen, late-nineteenth-century Americans cherished a totally stimulating surround in which not the tiniest space was wasted on emptiness. For them, studio style was perfect: it promised a home-décor project that was never finished but did not necessarily look unfinished, one that was a source of play, a scene of rapt visual attention, a discussable space, a site for experimentation with great ranges of colors and textures. Studio style, moreover, played to big-city views of space and its uses. Heaps of interesting and colorful goods were valued far beyond mere human ability to walk easily, rapidly, and cleanly through a room; free physical movement was sacrificed to a surround in which human beings took up positions to perform such concentrated activities as reading, letter-writing, conversing, and thinking. Here they could look up from their concentrated tasks, or outward from their thoughts, onto a range of visual stimuli that covered every surface, that moved in from the walls and swayed down from the ceiling and piled up from the floor. Here, undisturbed by bells or buzzers from the outside world, they were to find the promised *repose*.[8]

The prime American model for studio style was the grand studio held by William Merritt Chase in the Tenth Street Studio Building at 51 West Tenth Street. Chase's studio was famous, not only the biggest draw on Tenth Street but also a compelling model for anyone interested in accumulating and displaying goods imported from the global bazaar. In the opening scene of his 1894 novel *The Golden House*, Charles Dudley Warner conjured up Chase's studio—without Chase himself—for a midnight gathering of persons "under the impression that the end of the century is a time of license if not of decadence":

> The vast, dimly lighted apartment was itself mysterious, a temple of luxury quite as much as of art. Shadows lurked in the corners, the ribs of the roof were faintly outlined; on the sombre walls gleams of color, faces of loveliness and faces of

pain, studies all of a mood or a passion, bits of shining brass, reflections from lustred ware struggling out of obscurity; hangings from Fez or Tetuan, bits of embroidery, costumes in silk and velvet, still having the aroma of balls a hundred years ago, the faint perfume of a scented society of ladies and gallants; a skeleton scarcely less fantastic than the draped wooden model near it; heavy rugs of Daghestan and Persia, making the footfalls soundless on the floor. . . . Tacked here and there on walls and hangings, colored memoranda of Capri and of the north Woods, the armor of knights, trophies of small-arms, crossed swords of the Union and the confederacy, easels, paints, and palettes, and rows of canvases leaning against the wall—the studied litter, in short of a successful artist, whose surroundings contribute to the popular conception of his genius.[9]

William Merritt Chase was a dedicated shopper, a major consumer of fabrics, artgoods, and curiosities. He slid his shopping under double cover: his purchases had to do somehow with *art* and, arranged into crowd-pleasing display, with the promotion of art in America. William Merritt Chase, like Mrs. Bradley Martin and Edward Stokes, was exquisitely attuned to the iconography of desirable goods, to the objects people wanted to "inspect"; he thickened and clustered his displays of goods just as the Bradley Martins had. Moreover, Chase liked—though he did not limit himself to—several things Mrs. Bradley Martin favored: churchgoods transformed into décor, jewelry, costumes, and publicity. Unlike Mrs. Bradley Martin, Chase had no inheritance to fund his acquisitiveness. Chase was born in 1849 in Williamsburg, Indiana, where his father ran first a harness shop and then a general store; by 1861, the Chase family was in Indianapolis, where William Merritt Chase sold shoes at the New York Boot and Shoe Store. In 1872, a consortium of St. Louis businessmen put up $2,100 to send Chase abroad for two years of art study. In New York at the end of the century there persisted a vague "general feeling" that Chase "had money," but he did not; throughout his life, he earned his living by teaching, lecturing, and portrait-painting. He was frequently in the hole—but when he was in the hole, he continued to shop and to buy.[10]

By the time Chase hit the scene, the onetime "art fraternity" of New York City was no longer. In 1877, *The Art Journal* noted that either artists had withdrawn from society or society had withdrawn from them: artists' opening their studios for receptions and visiting had gone out of fashion. The social rich handed art-reception tickets over to "servant-maids, who came bringing with them youthful heirs of the first families in quest of entertainment." John Moran, writing in *The Art Journal*, suggested that Americans "absorbed in the contemplation and eager pursuit of material benefit and gain" had transformed art into a mere source of amusement. In 1889, when Elizabeth Bisland toured New York artists' studios, she pronounced New York City artists to be socially invisible: "The literary men and women, the musicians and the actors, have all become part of the social life of New York, but the world of the studios is a different social planet." Bisland herself, however, reinforced that difference: in her studio descriptions, she deployed again and again the terms *odd, strange, peculiar,* and *queer*. She reserved her personal enthusiasm for artists who lived on Broadway and who were "more in touch with the gayer social life of New York," who "had the air of men about town," and who had shown "a talent for organizing festivities" at Newport.[11]

In the 1890s, American artists were required to be not so much social as decorative and entertaining: they were asked to decorate their studios with things that the rich would like to "inspect," and to offer amusement in the form of not art but live entertainment. Many artists answered to the demand—and successfully, according to *The Art Journal:*

> In its appointments, the New York studio is more elegant than that of the foreign artist. Some of the great French masters work in *ateliers* which, though artistic, would seem quite unfurnished among the studios here, where there are frequently rich appurtenances—Oriental rugs, embroidered portières, carvings, and *vases ceramographiques*. The arrangement of studios is generally made with vast painstaking, frequently even with such cost of study as goes to the making of pictures.

Art critics uneasy with the situation wondered what, if anything, studio décor had to do with art. Edmund Tarver wanted to talk about art when he visited a studio; he complained that examining a studio's "picturesque appointments" both concealed and detracted from art. John Moran, however, averred that an artist had no choice: "Quite a large number of painters accumulate bric-à-brac, hangings, and other objects in their studios, not because they love or need them, but because it is the correct thing to do, and they must be 'artistic' or nothing."[12]

In 1879, John Moran visited the studio of William Merritt Chase in order to examine the *artistic life* for *The Art Journal*. Although Chase had been in residence but a year, he was already far gone into both creating and meeting the demands of the time for goods. His studio emerged from Moran's description as a museum without labels, a shop without clerks, a costume ball with but a very few revelers. The studio held, at that moment in 1879, the following.

Furnishings: carved Renaissance chest, sofa, tables, Venetian lamps, German silver lamp, and Nuremberg chair, its back "carved in grotesque semblance of a human face"

Churchgoods: an organ, a stained-glass window c. 1600 from a church in Northern Germany, prayer books, brass-mounted ecclesiastical records, old church velvets, an iron churchstand, wood carvings of saints and Virgins, and crucifixes

Japanese umbrellas, idols, and robes

Rugs, Venetian tapestries, handmade brocades, the yellow sail from a Venetian fishing boat, and a Spanish donkey blanket

Weapons: arms, casques, guns, pistols, and an Italian court sword

Musical instruments: bugles, drums, and tom-toms

Brass goods: a Persian incense lamp, a Turkish coffeepot, a doorknocker, candlesticks, and a lantern

Taxidermy: a stuffed raven, three stuffed cockatoos, and the head of a polar bear

Art: pictures and studies in antique frames, a bust of Voltaire, a head of Apollo, and a bronze medallion portrait

Old books, including a complete Cicero, quaint jars and Egyptian
pots, a facsimile of a Puritan hat, gourds, and an Italian coat of
arms

A collection of women's footgear, including "dainty little slippers
that have graced the feet of some sultana or favorite of the harem"

A palette and some paintbrushes

No hodgepodge effect here, however, and no embarrassing questions about the artist as mad consumer—not when William Merritt
Chase seated himself before an organ lodged in an alcove and bathed
his collection in an ineffable musical mystique. "A solemn, almost religious feeling comes over one," wrote Moran, "when with church
draperies and church lamps and burning incense around him, he sits
in the subdued light and hears the organ sounding from above." Chase
had Moran in the palm of his hand. A decade later, when the art writer
Elizabeth Bisland visited the studio, it was apparent that in the interval
Chase had not only acquired more of everything Moran had seen but
had also amassed new groupings of armor, altars, stringed instruments, and tapestry fragments. Moreover, he had put on display the
most memorable object he would ever acquire: "the glittering ebon
countenance of a prognathous Peruvian mummy suspended by its
long black hair." Looking back on the studio from the distance of
1916, the artist Candace Wheeler remembered, out of Chase's heaps
of bric-à-brac, only the mummy:

> There was a horrible little human head, no bigger than an
> orange and brown with the transparent brownness of the
> dark races. Some ingenious savage had extracted the skull
> and bones and had inserted bits of dark-red bottle glass for
> eyes, carefully preserving the long black hair and scanty
> beard, and with devilish dexterity shaping and smoothing the
> features until it was a fearsome shred of humanity. Mr. Chase
> delighted in it.[13]

In the presence of Chase's international bazaar, Elizabeth Bisland
did not mention Chase's priestly role, and she abandoned the original

fiction that his bric-à-brac functioned as props for his paintings: "They are rarely transferred to canvas," she wrote, "but serve to foster the owner's passion for color, and to give him the inspiration he draws in from the sense of dignity and sumptuousness in his surroundings." At that moment she assigned to Chase's goods a *raison d'être* that twentieth-century decorators and designers would perpetuate. Then, in one further effort to elevate Chase from consumer to *artist,* she suggested that his studio was quite like "the surroundings that the old masters loved so much."

Who was William Merritt Chase, once he ceased to be the vaguely priestly figure he presented to Moran in 1879? Depending on which group of objects occupied the studio foreground, Chase's own metaphorical costume changed: he might be a warrior, or a scholar, or an aristocrat. He was of no particular nationality; he was cosmopolitan. Certainly he was a man, and the thick layer of dust atop his many objects said so. If the dust vanished, a visitor might infer that Chase had dusted, and that inference was intolerable; men did not dust. Furthermore, Chase was an artist, and *artist* was the territory of men. *Shopper* was, in a nation of consumers, supposedly the territory of women, or at least a territory to which only women were openly lured, and a territory often scorned for having no one but women as its occupants. In cultural myth, men were supposed to make money, and although men might spend money on large items, especially on the transportational, men were popularly supposed not to shop, not to do a lot of looking in a lot of places for a lot of goods—and definitely not to hunt for bargains.[14]

From 1880 onward, Chase enhanced the look of his studio with display animals—a white cockatoo, ten brilliant macaws, a Russian wolfhound—and with a succession of African-American male servants whom Chase costumed as "Nubian princes." With this stroke, Chase went Catharine Lorillard Wolfe's home on Madison Avenue and Edward Stokes's Hoffman House Art Gallery one better. Their "Nubians" were bronze or ceramic slaves; his was alive. Within the studio, Chase himself wore the fez and lounged "on a couch under a Moroccan tent." When he emerged to stroll Fifth Avenue, he costumed himself as a French artist in a "black flat-brimmed French silk hat and a soft tie, and several rings from his growing collection," and on shopping expeditions he was accompanied by a wolfhound and the costumed "Nubian." When he played the role of professor, he wore "glasses with thin wire frames, a high stiff collar, and a cravat"; in the presence of the rich, he wore "a cutaway coat and trousers, a scarf drawn through a bejeweled ring securing his shirt, spats, a top hat, and a flower in his lapel." He began to entertain the wealthy with receptions and performances in the studio; like Mrs. Bradley Martin, he threw elaborate costume parties, and like her he fished for guests by sending out as many as a thousand invitations to a single reception.[15]

Regardless of his fluctuating situation, William Merritt Chase shopped and bought. He was one of those quintessential late-nineteenth-century American men—and they were all men—who lived in an agony of desire for more possessions and who could not put a stop to their buying. They eluded creditors by dodging into stores to buy more stuff. Like Stanford White, another colossal male shopper of the time, Chase was often in terrible financial difficulty, but most Americans who watched White and Chase believed both of them to be rich men. Chase's students, in contrast, accurately inferred the nature of his resources: they noticed that always, just after they paid their tuition, new objects appeared in the studio. On the subject of his shopping, Chase himself maintained silence.

In 1896, William Merritt Chase's spending habits caught up with him, and he was forced to list and sell every single object in the studio. Because Chase had successfully concealed his long-running financial difficulties, the sale announcement was a shocker. Chase's goods spoke not only of Chase himself but also of the culture in which he owned them; in truth, the possessor of an interior and its furnishings controls neither their accumulated cultural meanings nor even their personal meanings. Moreover, onlookers were awed and almost terrified by revelations of how much stuff Chase had accumulated. They could scarcely imagine where in the confines of a studio familiar to them he had stowed eighteen hundred objects, including:

Six hundred finger rings: signet rings, papal rings, poison rings, charm rings, betrothal rings, marriage rings

Pots, bottles, and dishes: Phoenician glassware, Bohemian ware, Persian ware, Roman goblets, Delft ware, Venetian jugs, Mexican vases, beer mugs, long-necked pitchers, ginger jars

Thirty-one musical instruments: fiddles from China and Japan, African tom-toms, Spanish and Italian mandolins

Weapons: silver-mounted dueling pistols, Spanish knives, old Roman swords, swords from Germany, Turkey, and Java, flint-locks, blunderbusses, trappings of Indian warfare, and a Zulu war shield

Lamps and lanterns from Venice, Persia, and Morocco

Fifty-five locks and keys

Japanese goods: glass teapots, marionettes, masks, and rice spoons

Thirty-seven plaques from Spain, Persia, and Morocco

Leather goods: Spanish bridles, saddles, saddlebags, and numerous other horsey trappings

Brass goods: pots, kettles, and thirty-five candlesticks

Two dozen pairs of women's shoes

Stuffed birds

One swan

Two ibises

One pelican

One raven

Fifty-eight tapestries

Furnishings: carved cabinets, Spanish armoires, long boxes, Dutch clocks, hall timepieces, and picture frames

Fabrics: cushions, costumes, draperies, hangings, and rugs

Jewel boxes

Thirty-seven Russian samovars

One Peruvian-Indian mummified head[16]

In their multiplicity lies their meaning. Between 1879 and 1896, the number of items in Chase's studio that suggested religious practice appears to have diminished, while his collection of taxidermy and pelts remained relatively stable. The collections that enlarged fantastically—weapons, musical instruments, smoking paraphernalia, rugs, brass wares, rings, and horsey goods—all point in just one direction:

Orientalism. Chase's full Orientalist collection was identical with the vision painted onto hundreds of French Orientalist canvases of the time, and identical to the objects that appear throughout a favorite text of the American nineteenth century, *Alf Layla Wa Layla,* or *The Thousand and One Nights.* It is here that Chase's studio appears conventional and here that it reveals him as a quite ordinary American consumer in every way except the gargantuan size of his shopping appetite. He was attuned to an array of goods, stories, and images with Middle Eastern and North African sources, items interchangeably labeled Persian, Algerian, Turkish, Oriental, and Eastern. He pointed to Orientalism when he donned the fez, lounged in his Moroccan tent, dressed up his American black servant as a real live "Nubian," and required him to appear as such both in the studio and on the street.

The nineteenth-century Orientalist passion crossed comparatively late to an America without wars or colonies to provide exposure to the Middle East. When Orientalism did arrive, after 1876, it hit America very hard. In the cultural cross, the Orient lost its people, except for a small army of *danse du ventre* performers, and arrived as goods, as a way to fill space, loosen up stiff décor, and suggest the life of the senses. Within a year of the Philadelphia Centennial Exposition's display of Orientalist goods, "sale-rooms were crowded to their utmost capacity wherever Eastern goods form the attraction," according to *The Art Journal,* which also thought that, typical of America, there was "something of excess in public appreciation of the Eastern productions just at this time; it seems likely to end badly."[17]

Across the nineteenth century, according to art historian Christine Peltre, "no form of expression was impervious to this incremental 'unveiling' of the Oriental, which in the nineteenth century affected everything from music to architecture, from literature to the decorative arts. In painting, this elsewhere—sometimes called 'the East' and sometimes 'the Orient'—was a point of convergence of all Western schools." Although under political and critical attack from the 1870s onward, Orientalist painting sailed on straight through to the end of the century, by which point everyone, whether traveled or untraveled, was producing it. In *Trilby,* George Du Maurier used two British painters, the Laird and Taffy, to satirize Orientalist fakery and its buying audience. The Laird specialized in toreador pictures that sold fast until he went to Seville and saw his subject, whereon his Spanish pictures "quite ceased to please." He moved on to Algeria, where he successfully painted Roman cardinals. Meanwhile Taffy, who joined him there, painted Algerians "as they really were" and was unable to make a sale. By the end of the century, and in response to sheer demand, Orientalist art became a standardized stream of harems, rug markets, and mosque scenes. No matter what the scene, however, the painting had a sort of shopper's corner, a nicely arranged grouping of the Orientalist goods desired in the West. To twentieth-century eyes, "serious" Orientalism and fake Orientalism are now nearly indistinguishable: in 1950, art expert Maurice Rheims dismissed all of it as "outdated and grotesque fantasies in which trumpery Arabs smoke hookahs under the eyes of unveiled creatures nourished on Oriental philtres and potions."[18]

Nineteenth-century art-writers addressing American audiences, however, drooled over Orientalism's potent combination of girls, goods, carpets, and color. Mary Cecil Hay told Americans about Frederick Leighton's *Light of the Harem*, in which "a beautiful Circassian favorite stands winding a scarf around her while a child holds the mirror for her." Lucy Hooper described Cabanel's *Vashti*, painted to the order of William Schaus of New York City, wherein a "shadow of encircling draperies," jewels, gold, ornamented chairs, and feather fans surround a life-size "haughty Persian queen." Her face is in shadow but her "robe of white gauze wrought with gold is open in front so as to show her beautifully molded throat" (the latter term is late-nineteenth-century code for breasts). Jean-Léon Gérôme's *Sword-Dance*, exhibited in New York City, offered a "loosely draped dancing girl" surrounded by flutes, guitars, tambourines, Moorish arches, divans, cushions, and coffee cups, all of it pronounced "exceedingly true to and characteristic of Eastern life." Americans could read further about the Oriental look, *sans* loosely draped women, in *Artistic Houses*, have it styled for them by Louis Comfort Tiffany and other less expensive decorators, experience it in the Waldorf's shadowy

Moorish Room, see it *en masse* in the "great Oriental bazaar of Vantine & Company" at 877 Broadway, and engender its fantasies in apartments like Jim Brady's, "furnished in the pseudo-Oriental style believed to have an aphrodisiac effect on impressionable young women." Some Orientalist painters sold well in America, but for sheer popularity they never touched the French Saloniste painters of nymphs and peasants: Orientalism hit America not so much through French Orientalist art itself as through the American effort to reproduce the interiors of French Orientalist art.[19]

Orientalist images and goods were, and still are, both protected by and concealed behind the word *exotic*, a term that slams the door on knowledge. In 1877, John Moran proposed that days spent exploring William Merritt Chase's "exotic studio" might allow him to "enter on a knowledge of its contents," but in fact Chase's studio did not offer knowledge. His was the Occident's Orient, where "exotic" meant "unknowably foreign." The hot term *exotic*, however, meant much more than merely foreign, especially when it appeared in harness with *secret, intoxicating, dangerous, mysterious, spicy*, and *titillating*, and when Westerners were said to have "succumbed to the lure of the exotic." In America, no Northern or Western culture, no matter how unfamiliar its foods and habits, was ever considered exotic; the exotic was located in an equatorial East from which it signaled a life that was lush and voluptuous—and also passive, cruel, and fatalistic. In the American consumer arena, *exotic* meant lavish displays of fabrics, artgoods, jewels, and collections that one would not be required to explain.[20]

The exotic global bazaar through which Americans rummaged at the end of the nineteenth century was ideal for consumers whose commitment was always temporary and who wanted to raid the cultural display but leave the cultures behind. Furthermore, American demand redefined the Oriental bazaar exactly as it had redefined Salon art: American importers such as Louis Comfort Tiffany not only sent agents to the Middle East to locate goods for American markets, but also ordered by the gross specific items desired in, and designed for, America. In going for Orientalist décor, Americans went for a lot more than mere portability, the loosening-up of stiff décor, and exciting shopping experiences. Along with the objects came the stories; every Western culture in the nineteenth century was saturated with the stories of *The Thousand and One Nights*.[21]

The Thousand and One Nights is a title more than it is a text, and as a title alone it could suggest adventure, violence, mutilation, horses, gold, jewels, shopping, easy money, rugs, lamps, flagellation, and sex with multiple female partners. There is no "original" text: compilers, translators, and expurgators selected from a mass of stories and poems when they built their versions of the *Nights*, and some selections appeared in every nineteenth-century library. A man might peruse the version to which Richard Burton had added his own racy and explicit material while in the same room children read the version that Kate Douglas Wiggin had framed with "moral lessons." In an America that divided appropriate reading along gender and age lines, the *Nights* functioned as a universal referent, even though no one could guess which version an auditor might be familiar with.[22]

A full Western collection of Orientalist objects, like that displayed in the studio of William Merritt Chase, shifted in shape and suggestibility depending on which *Nights* lurked in the psyche of its individual observers. Chase's collection of musical instruments, for example, spoke to a reader familiar with the stories of the thirty-third and thirty-fourth nights, wherein music produced by those same instruments lures Harun al-Rashid into a drunken party that eventually becomes a flagellation scene. Chase's Moroccan-draped lounge spoke to a reader who knew that from behind just such draperies in the *Nights* there usually emerged "a dazzling girl" with "sugared lips and Babylonian eyes," a girl who brought to mind "a dome of gold, or an unveiled bride, or a morsel of luscious fat in a bowl of milk soup." Chase's collection of locks and finger rings spoke to anyone who remembered the Prologue story that propels the rest of the text: a woman imprisoned in a quadruple-locked glass chest manages, with the aid of a demon, to escape the chest and have sex with a hundred men. From each man she extracts a ring as payment. Discovery of the ring collection and its meaning launches Shahryar on the serial-killing spree from which Scheherazade saves herself by telling stories whose conclusions Shahryar must wait to hear. Kate Douglas Wiggin excised that frame story from her children's version of the *Nights*.[23]

In America, Orientalist goods and customs spoke a language of bodies and sexuality that the dominant language itself could not speak in public—except to deny. One path to the Orient wound through the pages of the *New York Herald*, always attuned to the latest, but that path no sooner opened than it closed. The *Herald* was simultaneously daring and timid; again and again it suggested the *exotic* and then at once delivered a sharp slap to anyone hoping for same. The *Herald* edged toward the East in May of 1893 when it discovered Turkish baths and announced that "shapely women doted" on these "novel, refreshing, wholesome, appetizing, and eminently decorous" experiences. After further noting that the "shapely women" were consuming "drinks uncountable," the *Herald* suddenly wagged its finger at hopeful readers and announced, "If any one anticipates a recital of some wild orgy let him read no further." Then the story took a quick detour into the safety of Salon art: "If a new Bouguereau could paint the guests at one of these symposia the old familiar nymphs of the Hoffman House Café would find their luster dimmed by the contrast." Men also sought the Orientalist experience, according to a shampooer at a Turkish bath in Manhattan, but here the *Herald*'s treatment veered off into standard discussion of big-city money:

> Perfumed baths for men are no longer considered luxuries, but actual necessities. Men all want the best that can be bought, and the greatest luxury after a Turkish bath is to be rubbed with a delicious mixture of alcohol and some inoffensive perfume. It may truly be said that this is the day of luxury. It is the Sybaritic age. Men spend more money for the good things of life now than ever before. Call New York a billion dollar city, for never in the history of America has so much money been spent for luxuries and entertainments.[24]

In 1895, a new path to the East opened when the *Herald* reported on Oriental smoking rooms in Manhattan, where the Western clientele experimented with the narghile amid "divans upholstered with soft stuff, and covered with Turkish rugs and easy cushions, and little ebony tables and stands." Western sitting rules, however, still obtained: in the Oriental smoking room, the *Herald* warned, "the habitués DO NOT sit Turkish fashion." As nice Christmas gifts for a woman smoker, the *Herald* suggested a Turkish fez which would look "particularly jaunty when worn with a Japanese kimono," a narghile filled with rose water, a jeweled anklet—or a dagger. In 1893, these daggers, "superb objects, jewelled and golden, sharp and deadly," became a popular gift for young women, upon whose "boudoir tables they were allowed to repose." Orientalism knew the dagger: in that world, it was prostitutes who could be identified by their habit of carrying little daggers. The objects, as always, spoke to those in the know and were otherwise silent though not impotent. It was impossible to know who was catching which message.[25]

For children, by the end of the century, the door to the East both opened and closed with bowdlerized versions of *The Thousand and One Nights* and with such "entertaining and instructive" board games as Aladdin and Romany. For all the rest of America, the path to the East—first glimpsed at the 1876 Centennial Exposition—tangled itself in the long American fascination with Paris studio life, passed through French Orientalist art, took the stage in tinseled *Ali Baba* extravaganzas that linked "scenes of a fanciful Orient with the topical hits of present-day political life," flowered in Chase's studio and dozens like it, came to life at the Columbian Exposition's Midway *Plaisance*, crossed some version of *Alf Layla Wa Layla*, and ended in a draped and cushioned segment of anyone's living room called a *cozy corner*.[26]

News of the cozy corner flashed across America in newspaper and magazine articles, decorating manuals, and advertising. The cozy corner became a "mania" and spread "like a contagion over the entire land." Artists, whose authority as décor arbiters would never be higher, "pronounced in favor of it." In its basic form, the cozy corner consisted of a low divan set diagonally across one corner of the living room; the divan was draped with an Oriental rug and heaped with pillows. Suspended above it were "Baghdad curtains," which could be looped back on either side. Like *The Thousand and One Nights*, the cozy corner could be sensualized with incense and dim lighting; alter-

Corner of G M Allen's Room

natively, application of lace curtains and ferns could bowdlerize it into dull respectability. Its effects could be had cheap: according to *The Ladies' Home Journal*, a cozy corner could be "fitted up at a small expense" with shrewd deployment of packing boxes, figured denim or jute, an old quilt, and hair (for stuffing) purchased by the pound. The *New York Herald*'s flat-furnishing contest of November 1893 yielded a winner in a "cozy corner that contains a couple of divans, simple frameworks on which rest the softest of springs and mattresses covered with old rose velours and piled with multicolored cushions. The shelves above are all attached with screws so that they can be removed when some dreadful first of May [general moving day in New York City] shall make it necessary. The draperies are imitation Bagdad, so fastened to the ceiling by means of furniture rings that they can be removed to the roof and shaken now and then."[27]

That cozy corner was very simple; others were elaborated with unlimited decorative effects that left no surface untouched and no effect unsought. Another prize-winning plan for a cozy corner, this one called "very Oriental and artistic," was so powerful that it reached out from its corner and consumed the entire room. The directions for achieving it were written up for the *Herald* by its winning dreamer:

> Cover the floor with Turkish rugs of brilliant colors and design. In one corner have a divan two feet wide. This can be made like a plain bench with mattress and covered with a soft Turkish rug. Pillows of different sizes should go with this, covered with silk in oriental blues and reds, and also with the Turkish embroidered squares that come for pillows. At one side place a Cairo stand with brass tray holding Turkish coffee pot. This can serve as a smoking table as well and will give an artistic finish to the whole room. In the corner over the divan suspend from the ceiling a Moorish lamp of carved brass. A candle placed inside of this when lighted will give an Arabian effect, very picturesque. On a good sized reading table in the center of the room have an ornamental lamp of wrought iron with yellow china shade and base. An easy reading chair of leather should stand near this. The other chairs can be of rattan with two or more dark wood ones mixed in. Those in colonial style are the most artistic with claw feet. The seat of these should be upholstered with Turkish coverings. Cover the windows with Syrian cotton curtains and drape a Turkish curtain or portière over these. Drape the mantel with two India cotton tablecloths. The wall above the mantel should be draped with a pair of curtains of same material. To carry out this idea of a frieze, use goods that come by the yard in one foot width. It is handsome to tack them on the wall, just beneath the ceiling; other drapings in Japanese or Turkish embroideries hung on the walls make a rare and artistic room.[28]

The Orientalist cozy corner, however, seemed to have a life of its own that it did not open on demand to Westerners. William Merritt Chase had himself photographed on his "Moroccan-draped lounge," but most other Americans had their cozy corners photographed unpeopled. Who, after all, were Americans pretending to be when they settled into a cushioned and backless cozy corner for a spell of languorous lolling? Did an American woman keeping house in an American city occasionally topple over into her cozy corner and become the idle and bejeweled odalisque featured in the tackier Orientalist paintings, or did she become a "King's Favorite" out of the peepshow books, waiting only to offer herself to the sultan? And who was the sultan? The cozy corner was certainly not social. Two persons lounging amid cushions on an unconstructed divan under looping draperies felt awkwardly disconnected from two persons sitting across the room in wing-back chairs, and the occupants of the chairs felt as if they were the audience for some performance that refused to begin and that perhaps they did not care to see anyway. The *Herald*'s busy décor writer tried to hand off the cozy corner to "lovers or newly married people," suggesting again its ineradicable sexual aura, but even that positioning turned out to be temporary: "Tête-à-têtes are very delightful for awhile," the *Herald* opined, "but they become fearfully monotonous." Finally, desperate to banish a style once thought "too obviously delightful," the *Herald* pronounced the cozy corner to be a "barbarous fashion."[29]

Across 1896 and 1897 the great interior-decorating binge that had occupied most of the nineteenth century in America lost its energy. Chaos had been sighted, and in the face of it, the powerful old arbiters of taste chose silence. Interior decorating articles became a rarity in the newspapers, and in 1898 *The Decorator and Furnisher*— published since 1882—shut down. Even the *Herald* seemed to tire of interior décor: it had glimpsed "the folly of over-ornamentation" and confessed to having possibly "run things into the ground." It seemed clear that "all sorts of old trash, gilded or bronzed and hung on the walls, did not produce the same results as anticipated." Now, said the *Herald*, it was time to "go in for solidity and comfort," and possibly time to redo the domestic landscape in blue and white with wicker furniture. The binge was not quite over, however, for in its very next breath the *Herald* said, "When once you get the blue and white craze you will SIMPLY NOT KNOW WHERE TO STOP." Meanwhile the Orientalized studio with cozy corner moved uptown in the hands of painters engaged in the "society racket." Male artists offered their clients luncheons, teas, dances, and suppers in business-entertainment spaces that the *Herald* defined as "entirely apart from art." Shoppers turned away from the Orientalist look: in February of 1897, after an auction of five hundred lots of Oriental carpetings and draperies, the expert auctioneer James Silo characterized the sale as "funereal. I have never had a finer lot to offer and have never sold goods of this kind so cheap."[30]

So it was at exactly the worst moment—for him—in the history of American stuff that William Merritt Chase was forced to sell the contents of his studio. In an effort to conceal his financial distress, Chase made a great mistake: he announced that he was giving up the studio and its contents because he wished only to paint. Americans, as Chase should have remembered at that moment, wanted to *buy into art*, not to buy goods discarded in favor of art, goods that were apparently obstacles to art. Chase's explanation looked like a bald admission that the famous studio had not, after all, been about art. And so, as might have been expected, out came the bargain-hunters.[31]

At the American Art Galleries, across four days in January 1896, eighteen hundred lots of Chase's goods were auctioned by Thomas Kirby, the same expert who had impressed John Tod at the sale of A. T. Stewart's art collection in 1887. Kirby was fast, witty, and adept at "prodding lazy bidders and bringing out unexpected offers." Although the saleroom was full, the best seats were occupied by auction-loving nonbidders who, like the onlookers at Cornelia Martin's wedding and the Bradley Martin ball, wanted only to see a show of goods. During the first two days of the sale, Chase's brasses, hangings, and *objets d'art* drew respectable bids. The Peruvian Indian mummified head, core element of the grotesque in Chase's collection, asserted itself again and elicited the most spirited bidding of the entire sale. On the third day, however, three hundred of Chase's finger rings did very poorly. The ring collection had been rather private, never displayed *en masse* in Chase's studio; at the auction, the rings made no show at all when each was described and offered singly. Eventually the rings realized less than half of what Chase had paid for them. It was the evening of that third day, however, that constituted real disaster: bidders displayed no interest whatever when sixty of Chase's own paintings came up for sale. Many of the pictures went for less than the price of their frames. Thomas Kirby told the *Tribune* that the bidding was "spiritless, and the prices entirely unsatisfactory." The *Times,* clearly embarrassed for Chase, characterized the sale of his pictures as a "keen disappointment" and appraised the sale overall as both "disheartening" and "disastrous." The *Tribune* noted that "it is a matter of considerable surprise among artists that Mr. Chase's own pictures did not bring higher prices." On the final day of the sale, the remaining three hundred finger rings went for next to nothing, "the dealers showing but mild interest, and in most cases getting the rings on the first bid." The four-day sale realized a total of $21,252.25.[32]

The delicacy of one's position on the American consumer landscape! Even consumption-drunk America had venues wherein persons were perpetually uneasy about selling: in the home and in the classic nonprofits of classroom, church, and museum. Putatively, Chase's studio, by his own making, had been all of those four. Chase's students had invested in his knowledge and expertise when they took his classes; they had not bought from him. Furthermore, even if onetime visitors to Chase's studio had admired there any number of his objects, he was not thereby well positioned to sell those objects to them, nor, if he did, could he expect to profit by the sale. In 1896 America, Chase looked like an accumulator, an American spender, someone who had overinvested in a stock that was on the downswing; there was a bear market in Orientalist studio goods. When Chase went under, as so many Americans far richer than he had done, American shoppers showed up to inspect goods whose value was diminishing fast. The goods were not art; they were only merchandise.

NOTES

1. In the 1880s and 1890s, *bric-à-brac* was a standard and nonpejorative term for curious, antique, and artistic objects. The term puzzled French visi-

tors to the United States. In 1895, Guy de Soissons wondered why objects that in France he called *bibelots* were called in America, "without any reason, *bric-à-brac*" (81). When the twentieth century discarded the nineteenth century's stuff, *bric-à-brac* acquired negative connotations and came to signal random junk not deserving of individual notice.

2. *New York Herald,* 1 December 1895; *New York Times,* 1 December 1895.

3. *New York Herald,* 17 September 1893; Muirhead, 30. In the depressed economy of the mid-1890s, big-city America went auction-mad, and retailers moved fast to take advantage of the mania. A bill calculated to suppress "fake" auctions of the desirable stuff of the time went to the New York legislature when auctioneers began, according to the *Tribune,* "to buy up the stocks of dealers who are heavily involved, paying a sum less than one-third of the value of the goods, and to sell the goods at auction at a large profit, leaving the creditors of the dealer entirely in the lurch. Another class of auctioneers is selling goods of inferior quality, claiming that they were imported or are of Oriental workmanship" (*New York Tribune,* 9 January 1896). Rich men like George and Edwin Gould sent their mothers-in-law to inspect the goods of the rich, and had agents bid for them at auctions like that of Mrs. Albert Flake of 184 Riverside Drive, a shopper who had bought a great deal at the sales of other rich people's goods: she had "bronzes and paintings from the collections of A. T. Stewart, George I. Seney, and Mary T. Morgan. The dining room furniture was formerly the property of William Waldorf Astor. Other curiosities were an ormolu clock that once belonged to King Ludwig and a writing desk that was the property of Marie Antoinette" (*New York Herald,* 14 April 1899).

4. Sheldon, I, 26–33. See Charlotte Gere, *Nineteenth Century Decoration,* for a full history of the century's décor efforts.

5. *New York Herald,* 17 December 1893; *Art Journal,* 1877, 329ff., 361ff. How hard it was to get rid of stuff is demonstrated by the following *Herald* story of 29 December 1895: "Much of the furniture which has abounded for the last decade or two—the glued together, rickety, so-called Louis Quatorze and Louis Quinze and Louis Seize abominations—were not built for a ripe and honorable old age, and after the machine made wreaths and scrolls and renaissance ornamentations have let loose a few times and been plastered on upside down by well meaning but inartistic repairers, we are glad enough to dispose of our lares and penates, but how? We cannot give them to the poor because, like our cast-off party gowns, they are unserviceable and unsuitable. So the second hand dealer is called in to help us in our emergency, and under the skilful manipulation of that worthy they are soon sold for more than their

first cost to the aspiring but unwise wives of humble artisans, and we reflect too late that true charity would have consisted in cremating them."

6. *New York Herald,* 26 September 1893.

7. Cook, "Studio-Suggestions for Decoration," 237. From 1877 onward to the end of the century, the world of the *artistic* did nothing but enlarge. *The Art Journal* regularly included, in the catalogue of things potentially artistic and thus worthy of study, clothes, jewelry, crafts, interiors, exteriors, hotels, studios, pictures, workplaces, illustrations, napkin rings, linoleum, lace curtains, bookbinding, sideboards, inkstands, bells, book covers, picture frames, shoes, and hair arrangements and comb-overs.

8. Americans in the 1880s were expected to be carrying mental palettes that allowed them to visualize color differences described in black-and-white texts. George Sheldon's *Artistic Houses,* for example, required a reader to discriminate among twenty reds: bright red, subdued red, dull red, tawny red, orange red, brownish red, madder red, crimson, garnet, ruby, cherry, crushed-strawberry red, pure Vandyke red, Alhambra red, capuchin red, pale Venetian red, Pompeiian red, carnation red, Brandon red, and Indian red.

9. Warner, *The Golden House,* 1–3. In contrast to Warner's treatment, William Dean Howells in *A Hazard of New Fortunes* called up not Chase's studio but Chase himself, in the person of Wetmore the studio-art teacher. In the novel, Alma Leighton remarks that she is certain Wetmore laughs at his female students behind their backs. Later Wetmore himself pops up in a restaurant scene and launches out on "the futility of women generally going in for art. . . . I'm able to watch all their inspirations with perfect composure. I know just how soon it's going to end in nervous breakdown. Somebody ought to marry them all and put them out of their misery." When Mrs. Wetmore says, "You ought to be ashamed to take their money if that's what you think of them," Wetmore responds, "My dear, I have a wife to support" (118, 140). The scene is an interesting 1891 view of Chase/Wetmore's public explanation of his personal expenditures.

10. Gallati, 10–18.

11. "Artist-Life in New York," 57–58, 121–123; E.T.L., "Studio-Life in New York," 267–268; Tarver, 273–276; Moran, "Studio-Life in New York," 343; Bisland, "The Studios of New York," 20. Chase's studio continues to be a controversial subject. Nicolai Cikovsky conceals Chase's shopping behind the *artistic:* "Chase declared his artistic presence and avowed his and his age's artistic mission most fully through the studio," and later, "Studio-temples inspired devotion and allowed an esthetic transcendence of mundane reality, and were, as in Chase's case, often costly testimonial acts of faith in Art and Beauty" (2, 8). Keith L. Bryant takes a business approach to explaining

Chase's studio; in his view, Chase was not shopping but setting up a shop wherein goods other than its apparent contents were to be sold: "A grand space filled with antiques, tapestries, palms, paintings, and bric-à-brac provided the colors, textures, and objects he loved to paint and gave him a place in which to entertain the wealthy, the well-born, and his friends" (64–65). Annette Blaugrund in *The Tenth Street Studio Building* is less comfortable with Chase's goods. She characterizes his accumulation as "obsessive," notes that "nothing seemed too inconsequential for his taste," and characterizes the overall effect as "a blur." Nonetheless Blaugrund assigns artistic utility to Chase's goods: "Some items became props for paintings, others were for inspiration, and still others for aesthetic ambience—all served as subject and background for his paintings" (114). The latter claim did not gain contemporaneous belief. The Tenth Street Studio Building fell to the wrecker's ball in 1956; an apartment building at 45 West Tenth Street now occupies the site.

12. Tarver, 273–276; Moran, "Studio-Life in New York," 343. Big-city Americans were believed eager to annex goods previously owned by artists. The English painter Philip Burne-Jones, whose baronetcy only added to his cachet, wrote that when he was about to depart New York City, "an auctioneer called to see whether he could sell any furniture that I might be leaving behind, to be advertised in his sale list as the 'contents of my studio.' " Burne-Jones had nothing but a sofa and a "cheap bedstead" he had bought on Eighth Avenue. The auctioneer, however, assured Burne-Jones that the furniture would "suit his purposes," and he also bought "a few rough pen-and-ink scribbles that were lying on the table." A few days later the auctioneer advertised in the *Herald* a sale of "original drawings and the numerous and beautiful appointments of the studio of Sir Philip Burne-Jones." He also produced "a catalogue of eighty-one pages, with reproductions of antique armor and furniture," and at the sale "waxed eloquent over the heirlooms left behind, appealing to the audience when bidding was slack and saying that Sir Philip would weep if he could hear the price that his grandfather's helmet was going for! He made a seven days' sale out of it." Burne-Jones's landlord, also an artist, spent a chunk of his wife's fortune on fitting up an enormous studio with a "splendid organ played by an electric attachment, innumerable skilful arrangements for the subduing or intensifying of light, and a vast Eastern couch, suspended by gilded chains from the lofty ceiling, so that it swung luxuriously a few inches from the floor, as you reposed in Oriental *abandon* among its sumptuous cushions. Hard by were little Persian tables furnished with perennial supplies of cigars and cigarettes and whiskies and sodas." It was a mystery to Burne-Jones "how this man succeeded in painting at all under such distracting conditions" (240–242).

13. Moran, "Studio-Life in New York," 345; Bisland, 14; Wheeler, 244.

14. Sarah Burns, in *Inventing the Modern Artist*, further discusses efforts to firm up the male artist's identity as a manly one (160–168). In contrast to the intense decorative efforts of men artists, women artists conceived of their studios as workplaces. Visitors to their studios, however, seemed uninterested in the women as artists. When Elizabeth Bisland described the New York studios of Rosina Emmet Sherwood, Rhoda Holmes Nicholls, and Dora Wheeler, she kept her focus primarily on the presence of babies in their studios and thus positioned the artists as mothers (20). John Moran, however, directly expressed woman's two most powerful cultural connections to the studio—dust and sex—when he wrote, "There are lady-artists, and good ones at that, with cozy little studios which invariably by their neatness betray their feminine occupancy; but what sort of Bohemian orgies are there held it must be left to one of the sisterhood to betray" ("Artist-Life in New York," 123).

15. Bryant, 67–69; Pisano, 41. As to the fez: many "artistic" people wore the fez when indoors, and for more than one reason. In 1902, *Punch* caricaturist Harry Furniss went to Otto Sarony's New York studio to be photographed. While Sarony rubbed soot into Furniss's hair to make it look thicker for the camera, Furniss noticed that Sarony's own "luxurious head of hair was not a fixture. He wore a fez, and as he paused and pirouetted and struck attitudes, he would pull the fez over one eye coquettishly, or over the other one ferociously, and with it went his hair, parting and all" (60–61).

16. *New York Times*, 3 January 1896.

17. "Oriental Embroideries," 132.

18. Peltre, 10; Rheims, 133; Du Maurier, 145.

19. Mackenzie, 55; Hay, 218–221; "Art-Notes from Paris," 379; "Gérôme's Sword-Dance," 164; Kendall, 35; Burke, 132.

20. Moran, "Studio-Life in New York," 343–345; Verrier, preface; Poltimore and Hook, 14.

21. In *On Longing*, Susan Stewart points out that "the exotic object is to some degree dangerous, even 'hot.' Removed from its context, the exotic souvenir is a sign of survival—not its own survival, but the survival of the possessor outside his or her own context of familiarity. There is always the possibility that the object itself will take charge, awakening some dormant capacity for destruction" (148).

22. Kate Douglas Wiggin, in the preface to her 1909 "special version" of the *Nights*, explained that she shortened the stories, omitted "tedious repetitions," and left out "all the pieces that are suitable only for Arabs and old gentlemen," but at the same time "took no undue liberties." At about that same time, young Al Capone joined the Forty Thieves Juniors Gang. Robert Irwin's

The Arabian Nights: A Companion is an excellent guide through various versions of the *Nights*.

23. Mahdi, 77–80, 9–10, 69.

24. Day, I, 39–40.

25. *New York Herald,* 14 May 1893, 19 November 1893, 27 January 1895, 6 December 1896, 5 December 1893.

26. *Chicago Tribune,* 4 June 1893, 25 June 1893.

27. *New York Herald,* 5 February 1894, 15 October 1893; Thomson, 27.

28. *New York Herald,* 19 November 1893.

29. *New York Herald,* 15 October 1893, 5 February 1894, 11 February 1894. In July 1894, Alexander McDonald, United States Minister to Persia, returned briefly for health reasons to New York and offered a class-based view of Persian women well attuned to Orientalist art: "Do women in Persia have a pleasant life? Indeed they do. They have great gardens to walk in and fountains to sit beside and get cool. They can drive over broad acres all their own—if they belong to the fortunate ones, as nearly all women there do—and they can stop and pick fruit and eat without thought to the future. Do women work there? Oh no; except those *born to work*" (*New York Herald,* 1 July 1894).

30. *New York Herald,* 29 December 1895, 10 February 1895, 27 February 1895. Artists engaged in the "society racket" included Stanley Middleton, A. Muller Ury, and J. Charles Arter. William Leach in *Land of Desire: Merchants, Power, and the Rise of a New American Culture*, discusses the eventual transfer of 1890s Orientalism to American popular fiction and film, where it has had an exceptionally long cultural life (104–111).

31. The actress Sadie Martinot made the same mistake when in May of 1894 she announced a three-day sale of her artistic goods and bric-à-brac. Martinot proposed to "free herself from the fascination of bric-à-brac, beautiful antiques, and other treasures, in order to devote herself more wholly to art." The disastrous sale yielded Martinot less than 10 percent of her original investment and "made the rough path of true art seem rougher still" (*New York Times,* 11 May 1895).

32. *New York Times* and *New York Tribune,* 3, 8, 9, 10, 11, 12, 19 January 1896.

"LIKE SCULPTOR'S WORK.

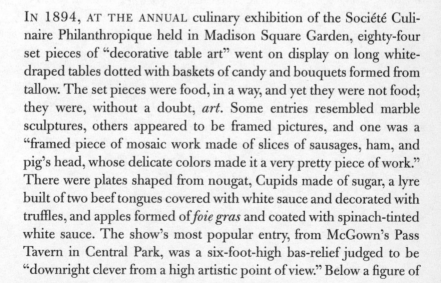

CHAPTER 4

Staring at Class

Fifth Avenue was a sight after church. Throngs emerged from the various sacred buildings and joined in the parade of fashion that moved at a snail's pace up and down the west side of the avenue. There was nothing here to indicate the city's poverty.
—*New York Herald*, 25 December 1893

IN 1894, AT THE ANNUAL culinary exhibition of the Société Culinaire Philanthropique held in Madison Square Garden, eighty-four set pieces of "decorative table art" went on display on long white-draped tables dotted with baskets of candy and bouquets formed from tallow. The set pieces were food, in a way, and yet they were not food; they were, without a doubt, *art*. Some entries resembled marble sculptures, others appeared to be framed pictures, and one was a "framed piece of mosaic work made of slices of sausages, ham, and pig's head, whose delicate colors made it a very pretty piece of work." There were plates shaped from nougat, Cupids made of sugar, a lyre built of two beef tongues covered with white sauce and decorated with truffles, and apples formed of *foie gras* and coated with spinach-tinted white sauce. The show's most popular entry, from McGown's Pass Tavern in Central Park, was a six-foot-high bas-relief judged to be "downright clever from a high artistic point of view." Below a figure of a "jolly little cook" there appeared, done in tallow and wax, four great end-of-the-century figures: a dancer in flounced tulle, a fashionably dressed man, "an anarchist in the very act of throwing a bomb, and with a cringing look of terror on his face," and "a wan and weary mother with her coaxing babies beside her." Too bad the bas-relief could not dramatize the predictable social intertwinings of its four figures: the fashionable man planning to seduce the dancer, the anarchist flinging the bomb at the man, the anarchist guiding the man on a slumming tour where he might glimpse the weary mother, the man leaving the weary mother at home while he went off to entertain himself with the dancer. . . . Visitors to the culinary exhibition apparently enjoyed the complicated fun in so artistic a union of social commentary, food, and illusion. The bas-relief was the most popular item in the show, and throughout the day crowds surged around it for a look.

Late-nineteenth-century Americans enjoyed a show of luxury goods, especially when the goods had been subjected to the decorative impulse. A lump of *foie gras* costumed as an apple was worth going to see. The edible made artistically inedible was just as pleasurable as the useful decorated into uselessness; tallow shaped into a social idea fell into the great and expansive category of entertaining decorative illusions, along with such other favorites as costume balls and cozy corners. The line between illusion and fraud was, however, mighty hard to draw, and as a result Americans who loved illusion were also Americans susceptible to fraud. They could be fascinated by a human being who projected an interesting and decorative illusion—by, specifically, a titled person enrobed in the illusion of hereditary privilege. Although some Americans were ready, on the purest republican grounds, to dismiss every title as a fraud, that ideology had a weak grip in the 1890s. To a great many Americans, royals and nobles were like the perishable *foie gras* apple and the tallow anarchist, worth going out to see.[1]

At the end of the nineteenth century, America's big cities were newly aswarm with persons who identified themselves as *sir, lady, count, countess, marquis, baron, baroness*, and *duke*. To most Americans, they were impressively unfamiliar people. Consequently, individuals "entitled" by the accident of birth could work the advantages of their titled positions, while others could bestow titles on themselves and thereby work the same advantages. All of them expected to move, so to speak, directly to the head of the line, where they further expected to be noticed, listened to, and catered to. They also expected to consume freely what America had to offer, to be treated and entertained by Americans, and to occupy the best rooms at the best hotels without necessarily paying. In sum, they asked that their titled illusion be honored in a country that constitutionally refused to honor their illusion.

Carrying a title through America could be both risky and unpredictable: title bearers had to remain fully garbed in their titled identities while at the same time accommodating themselves to a country without titles. Titled persons were, moreover, expected to enjoy displays of "democratic manners"—and to honor the idea of American equality—while they themselves acted out a display of hereditary privilege carefully made as inoffensive as possible to Americans. The lure of a hereditary title could easily draw an initial audience of Americans eager to inspect an individual who had apparently not been required to achieve, who needed only to be, and who lived high protected by nothing more than a single word. That initial enthusiasm, however, was fragile. Moreover, Americans contemptuous of titles generally chose to politely avoid their bearers; as a consequence, titled persons wandering through America met only those Americans who were briefly awestruck by them, fascinated by an identity inaccessible to Americans, and even enthralled by a person of whom effort had apparently not been required. On American soil, where "fake" titles were difficult to distinguish from "real" titles, all of them were troublesome.

In 1893, the World's Columbian Exposition lured to America titled personages intent on appearing in their national buildings on the Chicago fairgrounds and eager to be fêted for their very appearance. Their path to Chicago was not smooth. Trial balloons might have to be sent up before the visit of an especially questionable title: the proposed visit of Archduke Franz Ferdinand, heir presumptive to the throne of Austria-Hungary and understatedly described as "unpopular with the masses," was eventually canceled by Emperor Franz Josef when only Henry Villard, discredited railroad-wrecker, rose to display enthusiasm for the Archduke. For titled persons connected to the American mythology surrounding Christopher Columbus, however, the American fuss was huge. None elicited greater excitement than the Duc de Veragua, lineal descendant of Columbus himself, and the Infanta Eulalie of Spain, direct descendant of Ferdinand and Isabella. Eulalie sailed for New York on May 16, 1893, amid a diplomatic hubbub over whether President Cleveland would pay a return visit. Although differences over protocol were hastily smoothed over on both sides, the squabble was a preliminary signal of greater hubbub to come. The Infanta was about to cross cultures, and, like other European objects that had crossed to America, she would not be in America exactly the same thing she had been in Spain.[2]

Although welcoming American crowds limited themselves to an "unostentatious display of democratic simplicity" for her, Eulalie seemed pleased with her "rooms fit for royalty at the Savoy," and was further pleased when entire New York theater audiences rose to applaud at her entrance. She was "honored by a royal ball" at which "everyone believed that he or she said and did just the right thing when brought into contact with her royal highness." Eulalie's stay in New York City drew heavy newspaper coverage: fresh sketches of her accompanied daily reports of a drive she took, a breakfast she ate, her innumerable changes of wardrobe, and her jewels "so rich that Cleopatra might have looked on them with covetous eyes." Eulalie walked to church "in a democratic way," "accepted the compliments of new world society with simple dignity and grace," and laid flowers on the tomb of General Grant. So far so good for Eulalie; the social rich in New York City were honoring her for her title and she was honoring the manners of a country that did not honor titles. Only the *Chicago Tribune* reported that Eulalie went to Morris Park, bet on the ponies, and smoked cigarettes in the clubhouse. And an angry bubble surfaced from New York City's population of title-haters: Eulalie had been scheduled to review the Seventh Regiment in Central Park, but that event was moved to the plaza at the Fifth Avenue entrance to the park when unnamed Americans objected to use of the park for such a show. Nonetheless, the complicated and delicate illusion carried by and built for Eulalie held together until, on June 6, 1893, she departed for the Columbian Exposition in Chicago.[3]

The *New York Times* maintained an exceedingly mild interest in the city of Chicago but liked to launch little missiles at the Exposition. The *Times,* for example, gave considerable space to critics of the historical Columbus, editorialized on the "incredible number of deadheads—27,000 out of 92,000 per day—who contrived to make their way into the fair," and proved with statistics that most of New York City's beggars had departed for Chicago. Overall, the *Times* judged the White City's architecture to be "pseudo-Greek, principally pseudo," and warned Exposition visitors that they were likely to be scalded by heat radiating from those white buildings. In spite of such remarks, the *Chicago Tribune* perceived the *Times* to be the Exposition's only "sincere friend" in New York, unlike those newspapers that "laud the Exposition editorially but at the same time lend their 'news' columns to attacks on the Fair."

The *New York Herald* took no such divided position: it straightforwardly loathed both Chicago and the fair. To the *Herald*, Chicago was a mere pretender of a city, a self-proclaimed competitor unpleasantly different from New York; Chicago, moreover, had stolen a show that rightfully belonged to New York. Throughout 1893 the *Herald* kept up a drumroll of attack on Chicago's preparations for the Exposition: a "disastrous conflagration" was likely in firetrap hotels where a *Herald* reporter had seen paper walls being "hosed down" with cheap paint to conceal their defects. Three flimsy hotels, it was claimed, had already collapsed utterly under a single gust of wind. In another hotel a woman named Louise Bunker was found with "her brains hammered out" and "a false goatee," the sole clue, clutched in her hand. Her murder made it obvious to the *Herald* that Jack the Ripper had gone to Chicago for the Exposition. Chicago's streets were clogged with mud, the weather was vile, and the *Herald* had discovered that a visitor to that filthy city would have to pay five cents just to wash his hands. Chicago had built not just a fair but "a paradise for criminals," and planned no regulation of rampant vice; Mayor Carter Harrison "would accept the vote of the devil himself if he could get it and be glad of it." Moreover, Chicago had, in the *Herald*'s view, "declared war on New York": Chicagoans had rejected the work of New York artists, denied exhibit space to New York businesses, and refused to open club doors to visiting New York City clubmen. The *Herald*'s act was nothing if not complex. The newspaper, for example, raised the hideous specter of cholera and then asked fair managers—whose ignorance and carelessness *Herald* reporters had long since firmly established—for a denial. Then, when those men expressed "no great dread" of a cholera epidemic striking fair visitors, the *Herald* sneered at their denials.[4]

On June 5, two days before Eulalie's arrival in Chicago, the *Chicago Tribune* pronounced the Infanta to be a "most democratic woman"; in fact, the *Tribune* thought "it would be difficult to imagine anyone more thoroughly democratic." The Infanta and her husband, Prince Antoine, had, according to the *Tribune,* overcome the difficulties of being royal in America by "recognizing the fact that they must either render themselves amenable to republican institutions or suffer greatly in the estimation of the people of a democratic nation." New York society, said the *Tribune*, had failed to comprehend Eulalie's desire to be treated just like an American; the social rich had "made a ridiculous spectacle" of themselves in their attempts to "bow with great reverence, kiss the Infanta's hand, and get away without turning the back." In Chicago's welcome, by contrast, there would be no "silly display of ostentation," no toadying, and none of "that servility of flunkyism which has characterized another city."[5]

The newspapers of that unnamed "other city" continued to pay close attention to Eulalie as she moved west; the *New York Times* tended to report her official doings, while the *New York Herald* scrambled for inside stuff. When Eulalie arrived in Chicago on June 7, she was cheered by thousands on her way to the Palmer House, where she was lodged in the Moorish Suite and given a schedule crammed with breakfasts, dinners, receptions, and official tours. Just three days later she made her misstep. On June 10, Eulalie arrived late for a reception in her honor at the Potter Palmers' medieval mansion on Lake Shore Drive and demonstrated that she "was in no pleasant mood. She was introduced to several persons at the Palmers' home, but in the main did not acknowledge their homage by even so much as a bow. It was noticed that something was wrong but no one had any idea what it was. When the Princess had been at the house about an hour she and Mrs. Palmer were seen in conversation, and immediately afterward the Infanta left. There was consternation at the sudden departure of the royal guest." Eulalie's handlers put out a story that on her departure for the reception she—a personage for whose foot American sidewalks had been strewn with pansies—had been distressed to find not even a carpet covering the sidewalk in front of the hotel; she had retreated to her room, and had to be persuaded to go to the reception. The matter seemed very small, but who were Americans to say what loomed large for a royal personage? Three more days passed before the real source of Eulalie's "unpleasant mood" came to light in the *Herald:*

> The reason the Infanta Eulalie was not more cordial at the magnificent reception given at Mrs. Potter Palmer's home on Friday night was the discovery made by the Princess that the residence to which she had been invited was that of the man who owned and kept the hotel where she was staying. It was but a few hours before the time to depart when she discovered that the medieval castle on Lakeshore Drive she was going to and the tavern she was going from were both under the same management. Only then was the horrible discovery made that a daughter of Charlemagne and a Bourbon of Spain had been asked to drop around and visit people who kept an inn. The Infanta was at first disinclined to go to the reception, but was prevailed upon to do so when it was fully explained to her that Mrs. Palmer was the President of the Board of Lady Managers of the Exposition and in that capacity virtually represented the women of the United States. Upon this understanding the princess went, but it was whispered loudly in well-informed circles that she was not much pleased with the idea of accepting entertainment at the hands of an innkeeper, as Mr. Potter Palmer would be designated in the Infanta's own country.

The day after the reception, Eulalie arrived an hour late at a concert arranged for her by Bertha Palmer, and stayed but five minutes; Bertha Palmer, notably, did not appear at the concert at all. Later explanations of Eulalie's behavior only darkened the situation further. Early in her visit, said the *Herald,* Eulalie had announced her desire to experience "simple democratic manners" and was consequently disgusted by the "manner in which self-constituted social leaders aped the manners of the nobility, and fatigued by the demands of an incipient aristocracy, where she expected to meet an unaffected democracy." The *Chicago Tribune* did not answer the *Herald*'s claim that Chicago society had behaved exactly like the "servile flunkies" of New York, but its enthusiasm for Eulalie took a noticeable drop after the Potter Palmer reception. On the point of what it meant to be "a guest of the nation," Eulalie and her American hosts diverged completely. Eulalie, the *Tribune* reported, refused introductions, thwarted plans made for her, and stamped her foot when her whims were not immediately answered to. She liked reviewing troops, boating, and cornbread, but she demonstrated no acquaintance with the great American value of punctuality, and—perhaps worst of all—again and again she disappointed crowds assembled in the hope of "having a look" at her. She and the aspiring-to-be-royal Mrs. Bradley Martin made the same mistake: both of them failed to comprehend how much Americans would put up with just so long as they were allowed to *see* it. "Fun for the Infanta," the *Tribune* headlined, sourly implying that the fun was hers alone.[6]

When Eulalie left Chicago on June 13, no crowd and no Mrs. Palmer appeared at the railway station to bid her adieu; only Mayor Carter Harrison, apparently just as ready to deal with the devil as the *Herald* had claimed, appeared to say goodbye. A bitter *Chicago Tribune* editorialized that both "time and effort" had been wasted on the Infanta:

> Royalty at best is a troublesome customer for republicans to deal with and royalty of the Spanish sort is the most troublesome of all. The universal sentiment of those who have come in contact with its representatives will be that there were no common points of meeting between the Infanta and her attendants and those who were intrusted with the task of receiving them. In a word, they will feel that there is an insuperable chasm between the effete monarchism of Spain and the fresher

civilization of the New World. Some things that we did evidently were not appreciated by the Spanish guests. It is certain that several things done by the latter were distasteful and offensive to their American hosts. This being the case, it is better the two should part company politely but definitely, with mutual assurances that the acquaintance will never be renewed under similar conditions. . . . It was their custom to come late and go away early, leaving behind them the general regret that they had not come still later and gone away still earlier, or, better still perhaps, that they had not come at all. . . . These unfortunate occurrences have made the entertainment of royalty a burdensome, unsatisfactory, and humiliating piece of business and have inspired the hope that if any more royalty of the effete sort is coming here it will come incognito. In the free, breezy, bounding West, royalty is not appreciated or liked. Being in the habit of looking forward Americans do not get along well with those who look only backward.[7]

When she returned to New York City on June 15, Eulalie was escorted over "a double breadth of red velvet carpeting lined with potted plants and spread from the train to the exit in Forty second Street," but she did not return to the Savoy. Instead she went to a private house at Madison Avenue and Sixty-second Street placed at her disposal by a Spanish importer named Ceballos, and there displayed again her lack of democratic instinct; the *Times* noted without comment that at ten o'clock on the evening of the 15th, "the man servants who had come to America with the Infanta's party were standing on the front stoop of Ceballos' house. No quarters had been provided inside for them, and their chief spokesman, the valet of the Prince Antoine, said they had a good chance of having to stay on the street until morning." The discredited Eulalie then tossed overboard any claimed interest in "simple democratic manners" and announced that during the remainder of her visit she would travel incognita in order to escape the "American custom of indiscriminate introductions." When she visited Newport, her presence caused not a ripple, and in late June her departure from America got only a small notice on page 11 of the *Herald*. Meanwhile, newspapers across the country editorialized to the effect that "nobility in general and the Infanta in particular were not wanted in America, and that altogether too much was being made over the visitors."[8]

Eulalie's performance of royalty in Chicago had been altered by its American audience and then tangled in the difficult American relationship with royalty. Americans struck with a case of Eulalie fever did not, after all, imagine themselves to be somehow royal; that way lay madness, as in the case of Robert Garret, president of the B&O Railroad, who in middle life became convinced he was the Prince of Wales. To foster Garret's illusion his keepers hired "a whole staff of actors to impersonate gentlemen-in-waiting, court officials, Cabinet Ministers, and Ambassadors from other countries, and an expert was brought over from London to ensure that each of their costumes was correct." What Americans wanted with royal personages was a give-and-take, an exchange of performances: rich Americans wanted to dress in their best and perform a version of court manners, while Eulalie in return was expected to perform a version of the democratic. The two scripts, however, were hopelessly at odds: Americans gestured toward a royal Eulalie who was expected to gesture toward a version of America that she had imagined but that was not on display. Perhaps Eulalie's desire to meet "simple democratic manners" had meant that she believed class in America to be simple; in fact, to Eulalie, class in America was simply indecipherable.[9]

The Duc de Veragua, a lineal descendant of Christopher Columbus himself and a guest of the nation who was even more enthusiastically welcomed than Eulalie, also discovered that a title in America occupied an uncertain and wavering position subject to sudden drops and losses. On first arrival, Veragua was seen as somehow an American, or at least as someone who could be addressed as if he were an American. On the steps of the Waldorf, John Austin Stevens, Secretary of the Chamber of Commerce, greeted Veragua as "Mr. Columbus." The duke, however, failed to offer any significant thanks for elaborate entertainments given him at city expense, and, on a darker American note, he was tripped up on his visit by, he claimed, having "caught the American spirit of money-making." Just before leaving Europe, Veragua had thrown his entire personal fortune into speculation on the Paris Bourse "in the hope of acquiring great wealth rapidly," but as he moved across America he received daily bad news of his investments and "his mind was burdened with thoughts of his financial ruin." Furthermore, he had had to demonstrate through expenditure that he belonged to the social level of the persons escorting him: "the fact that he was being so royally treated in America brought with it its penalties, for he was compelled to spend a lot of money, and this he really could not afford."

Finally, in Chicago, Veragua confessed that he was broke: his creditors, he said, had seized his house in Madrid, and he appealed for funds he said were needed to rescue his wife and children from penury. Although there was something sweetly appropriate about Mr. Columbus reaching America and going bust, the Duc de Veragua was suddenly in a very bad spot; as a beggar he collided with the resistance of the rich to persons positioned as objects of charity. Talk of raising a fund to aid him was unenthusiastic, and when he returned to New York he was ignored. His press coverage stopped abruptly, and his American story trailed off into silence. After a long absence from the pages of New York City's newspapers, Eulalie surfaced just once more: in November of 1893, her proposed election as an honorary Daughter of the American Revolution raised such a ruckus among DAR members that the officers supporting her election resigned their posts in a huff.[10]

Many Americans were far more pleased and comfortable with royalty and nobility when they could simply buy a piece of the illusion. The Vanderbilt women and others bought "beautiful bedsteads which once belonged to dead and gone royalty"; many Americans in the 1890s visited Tiffany's department of Blazoning, Marshalling and Designing of Arms Complete, chose from a massive book whatever devices appealed to them, and had Tiffany work them up into a "coat of arms." The next step in consuming hereditary privilege, according to the *Herald,* was to "get an English oak table. With one of these for your library and perhaps another for your hall, you may, if you have bought yourself a brand new coat of arms, now use it without any glaring inconsistency. With an ancient suit of armor and spears by way of poles for hanging the draperies, you will come to feel that you belong to a very old family indeed." The noble illusion worked better as purchasable goods than it did when attached to human beings—who seemed invariably to disappoint even those ready to be dazzled by royalty.[11]

Many foreign visitors to America ridiculed the illusion of buyable nobility, but they also, no matter what their class level or their country of origin, stumbled over the subject of class in America. They struggled to discern who if anyone was at the top, got confused over where ten million dollars placed a man on the class ladder, denied that there was a ladder, claimed that the rich wished to raise up the poor, and counterclaimed that the rich held the poor in utter contempt. Furthermore, in the 1890s, these visitors attempted their class analyses in the face of a wealth-holding gap greater than any they had previously observed: in 1892, the Census Bureau estimated that 9 percent of the nation's families owned 71 percent of the nation's wealth. There were 4,047 millionaires in America, but white-collar workers were lucky to bring in two thousand a year, industrial workers earned no more than five hundred a year, and department-store saleswomen earned three hundred a year. At the lower employment levels, a worker who got sick lost his or her job. Homelessness and unemployment, after 1893, were epidemic.

Foreign observers who had no trouble recognizing desperate poverty in America were nonetheless unable to sort out the class levels stacked above it. In America class was not necessarily available to the ear, it was not static but mobile, and it was rarely discussed openly. Confusion was reciprocal. If, on the one hand, foreign observers failed to perceive American class distinctions, on the other hand Americans could be temporarily fooled when a stranger pressed on them the illusion of foreign nobility—the great illusion not only of being better but of having been better for a very long time.[12]

In America, the illusion of being better required either resources or the illusion of resources; the latter illusion was always temporary, but many persons both American and European employed it. A performer who could annex an "aristocratic" identity might buy a great deal before anyone discovered that he could not pay. On December 2, 1895, John Sloane, "a tall smooth-faced youth of aristocratic bearing and faultless dress," appeared at the desk of the Hotel Majestic at 115

Central Park West; Sloane had no baggage but he did carry the "air of a traveller to whom the question of expense was no consideration." He ordered the best suite in the house, dined sumptuously in his room, drank—on account—throughout an evening in the Majestic's café, ate an expensive breakfast, and ordered up a carriage. Then he went shopping—without money. His *modus operandi* involved ordering merchandise to be delivered to the Majestic and giving as a reference the name of the Majestic's manager. Late in the day, when Police Detective Hunt finally chased down Sloane to a store on Fulton Avenue, he was still shopping. After Sloane was locked up in the West Sixty-eighth Street station, several hotel detectives dropped by to see if he was the same aristocrat who had performed the same act at the Empire, the Marlborough, and the Brunswick. Detective Hunt, impressed by the impenetrability of Sloane's illusion, judged the man to be "of unsound mind."[13]

A noble illusion could be milked longer if the performer had luggage or prospects: Cecil Foster, "Lord Crewe of England," always arrived with several trunks and two pug dogs, and paid his bills promptly for a week before he began to freeload. Count Zidzislow Kormorowski of Austria dined out for months on Long Island with a tale of a soon-to-arrive inheritance with which he would buy an "immense stock farm" from one of the men competing to entertain him with wine and dinners and cigars. In a notable 1894 scandal, a young Englishman—never named because the newspapers chose to protect the social rich in whose houses he had stayed—supported his aristocratic illusion and led a life of balls, nights at the opera, and club memberships by simply stealing any loose cash he spotted in the homes and clubs of the rich. His compatriots in New York City forced him to flee before his thievery further undermined their own performances of social superiority.[14]

With neither a title nor foreign birth to lean on, persons who claimed American class privilege sprayed a variety of credentials over their victims. They claimed blood relation or intimate friendship with important people, they displayed business cards and letterheads on which their names were prominent, and they produced, always with shows of reluctance, photographs of "estates" soon to be theirs. The illusion of being a "better" American, however, required further support, and usually gained it from haughty behavior and good clothes. Charles Whittaker of Holyoke, Massachusetts, "a young man dressed in the height of fashion," successfully cashed rubber checks in New York for a full year by representing himself as George Armour, son of Chicago meatpacker Philip Armour. Whittaker's *modus operandi* involved his mailing a letter and two bad checks to "George Armour" at, for example, the Park Avenue Hotel. Then, a day or two later, Whittaker appeared, registered himself as George Armour, asked for his mail, opened it in full view of the desk clerk, spilled the checks onto the desk so the clerk could easily read their dollar sums, and asked that the clerk cash the smaller of the two bad checks. T. Willington Osborne, on the other hand, massed his credentials by telling Mrs. A. F. Adams of 13 West Twenty-fourth Street that he was a Bostonian and a graduate of Harvard, class of 1877. Osborne used distant governors to build his identity: he said he was an intimate friend of Governor Russell of Massachusetts, and also let on to Mrs. Adams that he himself preferred to be called "Governor" because of his striking likeness to ex-Governor Campbell of Ohio. Osborne promised to introduce Mrs. Adams to "a real countess" staying in the city, and he claimed "large interests" in theaters and theatrical companies. But the promised free tickets to shows never materialized, and the countess was always too busy to meet Mrs. Adams. When T. Willington Osborne "had the blues he borrowed Mrs. Adams's cash, and she says he had them often." When her cash was gone, Osborne seized her jewelry and vanished.[15]

Such city identity swindles were no more than ordinary; also familiar was the sponger who apparently wanted not money but identity itself. For three years a man went about New York City representing himself as Walter B. Lawrence. The real Walter B. Lawrence was a well-known society man, a member of the brokerage firm of Dick, Benedict & Lawrence at 30 Broadway, and the owner of "a magnificent country place" in Flushing. During the workday, Lawrence's alter ego would drop by the homes of Lawrence's friends and acquaintances in the city; if he found a woman at home, he told her that he was Walter B. Lawrence, that he had lost his pocketbook while playing tennis, and that he needed a dollar to get back to Flushing. He repeated this act three and four times a day. Though he could easily have requested more than a dollar, he never did, and sometimes he asked for only seventy-five cents. "I never was so bothered in my life over a matter," said the real Walter B. Lawrence. "The man seems conversant with all my habits and all the persons I know. When I went to the Charities Aid Society about this, they told me that it was a frequent occurrence, and that my case was similar to hundreds of others that they knew about."[16]

Americans were interested in disguises, but their interest lay in stripping them away, not in paying for their upkeep. Moreover, as the century drew to a close, they registered in their entertainments a distinct preference for costume over masquerade: costumed persons displayed identities while remaining recognizably themselves, but masked persons were up to no good. Throughout the dark years that followed the crash of 1893, persons longing to recover the delights of a lost prosperity began to mask themselves as buyers. They sought to gain neither money nor goods from their masquerade; they apparently wanted a moment's notice and the thrill of consumer importance that surrounded anyone ordering heaps of goods. None of them thought small. In June of 1893, James Smith, a milk-wagon driver who lived at 145 Wallace Street in Brooklyn, represented himself around Manhattan and Brooklyn as a buyer for a large dry-goods concern in Guthrie, Oklahoma—a locale he had never visited. Smith placed orders for great quantities of fur hats, gloves, and firearms, all to be shipped to Oklahoma. When he was caught, and was asked in court why he had ordered the goods, "he said that he guessed he was crazy. 'They filled me up with whiskey,' he said, in speaking of the firms he had defrauded, 'and then displayed their goods and asked me for orders, which of course I readily gave. They were delighted for the time being and so was I.' "

A person without resources who masked himself as a buyer could touch many more lives than James Smith did. In 1893, a well-dressed young man masked with a business card that read "Howard W. Ream, banker, 40 Wall Street, residence Hotel Waldorf" traveled to Bristol, Rhode Island, to visit the Herreshoffs, the world's top yacht builders.

Ream placed an order for a "twin screw steam yacht, 185 feet long by 22 and $^1/_2$ feet beam, its bottom to be plated with Tobin bronze in the same manner as the *Vigilant* was, and to be built with all modern improvements conducive to convenience and luxury." Not only were the Herreshoffs happy, but the entire town of Bristol bubbled with cheer until Ream failed to meet the first payment. Work stopped, orders for materials were countermanded, the contract was torn up, and the mechanics were again out of work. Howard W. Ream, if that was even his name, was a fraud, riding on little more than a business card.[17]

The illusion of ordering for oneself the largest and most expensive personal transportation available in 1893 amounted to pure *buying*, uncontrolled by any details of receiving or paying. For a fully detailed consumer plan unrelated to the self, no one beat the elaborate illusions fostered by "Dr. Emil Blum," a man of "fine military bearing and waxed mustaches" who always gave his address as the Plaza. In September of 1894, "Dr. Blum" visited the offices of four different railroads, and at each office, after tossing around such names as Drexel, Morgan, Depew, and Pullman, he announced grandly that seventeen officers and 230 soldiers from the Austro-Hungarian army were about to arrive on the steamship *Armenia*, and he wanted two trains readied to transport them across the continent to Vancouver. From there, he said, they would embark for China and spend the winter "training the Chinese army in modern methods of warfare." To feed them along the route, "Dr. Blum" ordered a carload of chicken sandwiches and a ton of pie. Eventually the *Armenia* arrived in port with its hold full of nothing but beans, onions, raisins, and prunes; meanwhile "Dr. Blum" vanished. West Shore passenger agent C. E. Lambert, who had "impressed every chicken residing along his line into service to make the chicken sandwiches" for the expedition, said, "I can't make out why on earth Dr. Blum should have invented that steamer and filled her full of those fellows. He was in my office half a dozen times, always with a busy air, showed me at least a quart of decorations he had received from various crowned heads, and reams of documents proving what an important person he was. He wasn't doing it to get free passes, he didn't ask me for a cent, and his proposition was not one that would have secured him anything from any railroad. What was his object?" Blum's moments of "delight" occurred when he consumed at a high and important level, saw railroads and chicken farmers and bakeries all mobilized to his command, moved illusory armies, and affected international relations.[18]

Sloane, Ream, and Blum all relied on the cut and quality of their clothing as a basis for their credibility. In an America where good clothes were persuasive, to have them and show them, say, at midday every Sunday on Fifth Avenue constituted a type of social position in itself. The British visitor Alexander Francis was

> impressed by the comparative absence of the outward and visible signs of deep and widespread poverty seen, by the most casual observer, in the dress of large classes of citizens in other lands. It seems as if Americans, alone of all peoples, had not the poor, at least in appreciable number, always with them. But things are not what they seem; and the appearance of Americans is a deceptive guide to their true estate. The poor are here. Their democratic training, however, has led them to presuppose that a common standard belongs to all; and to this standard they are constantly striving to conform. The familiar tokens of poverty are found in the homes of multitudes of those whom one sees elsewhere in brave attire. Poverty in America that seeks to hide itself beneath fine apparel may be more bitter and more dangerous even though it be less than the poverty that elsewhere is naked and unashamed.[19]

How much bitterness and danger lurking under fine apparel joined the regular Sunday midday parade on Fifth Avenue no one knew. "People of both sexes," according to the *Herald,* made it "a rule of life" to be seen in that parade even though "they don't live on Fifth Avenue and don't necessarily know anybody who does." The social rewards emerged, however, from the *Herald*'s eventual recognition of what clothes meant:

> Fifth Avenue is a great street and there are a lot of people in it. How easily to be mistaken for one of them! There is a world of satisfaction in the idea. On Sunday Fifth Avenue doesn't ride,

and anybody may join the procession—anybody who has
good clothes. For Fifth Avenue only insists on the matter of
good clothes. That is what the procession is for. There is really
no place in the world where good clothes can be shown to
such advantage as on Fifth Avenue on Sunday noon.

Americans strolling Fifth Avenue wanted, as the anthropologist
Mary Douglas puts it, to show others that they were "equipped at the
same level." Equipment, at the end of the nineteenth century, meant
clothes. The effort to show oneself as "equipped" has never failed to
draw sneers from those Americans who have taught themselves to dismiss consumption and display as trivial emulation and from those
who seek to position themselves as superior to such shows of consumer effort. To look down on any American's effort to be equipped
as well as others is, however, to set oneself not above that effort but
against it, on "the side of the excluders, and against those who are trying not to be excluded."[20]

To experience the "world of satisfaction" that, according to the
Herald, lay in being "mistaken for one of them," an individual had to
be seen among them, had to occupy the same venue that "they" occupied. The Hackensack meadows did not constitute one of those venues. In August of 1893, as the army of the unemployed grew in size
daily, a "natty crowd of men" took up residence on "an oasis in Jersey's swampy meadows at the mouth of the Bergen tunnel." Even
though railroad employees called them *tramps* and *thieves*, other
observers labeled them *commuters, dude tramps,* and *the genteel unemployed.* They looked like "men accustomed to work," and each day
they produced razors, combs, brushes, and pieces of soap, shook out
their clothes, and generally made themselves presentable. Theirs was
a group effort, but its results were purely local: young women from
Jersey City Heights regularly dropped by to flirt with them. The *Herald* reporter talked to one of them, a man who was "sitting on his carefully folded trousers while his coat was spread on the grass near by. He
had been freshly shaved and his face shone. He was a sturdy fellow of
twenty-eight. 'I didn't come here in a sleeping car,' he said, 'and I'm
trying to get my clothes in presentable shape. You see the crease has

gone out of my trousers.' " He would not give his name, but he knew that clothes were no mask, no veil over some "real self" that twentieth-century Americans would dream into existence. No, in the 1890s clothes *were* the self and every big-city person knew it.[21]

Within the context of clothes-power, an American who was not arrayed as "one of them" courted humiliation and rejection if he attempted to enter the venues of the well dressed. In January of 1897, when a plumber working on a Fifth Avenue house near Eighty-second Street decided to spend his lunch hour in the Metropolitan Museum of Art, he was denied admission because he was wearing overalls. The plumber protested that he was a citizen and had a right to see the pictures. D. W. Kellogg, the Met gatekeeper, was adamant in support of the Met's "unwritten rule that a man must be properly dressed, and his clothes must not be dirty, and he must have all his clothes on." Kellogg further announced that he was the judge of what constituted proper dress; then he summoned Park Policeman Michael Tomkins from his lunch in the basement of the museum and directed him to eject the plumber. Tomkins refused to do so, but he did walk the man outside; Tomkins said that he was simply preserving the peace and most definitely not carrying out the museum's "unwritten rule." Curator William B. Arnold, however, made the social situation hotter when he announced loftily that he himself would not admit to the Met a man in overalls, a man in shirtsleeves, or a woman with a shawl over her head. He would, however, consent to admit a man in bicycle costume. Other museum officials took other stands on the plumber's overalls: The museum's president, Henry G. Marquand, claimed unfamiliarity with the "minute laws" of the museum, but William L. Andrews of the executive committee opined that "one of the objects of the Museum is to educate and to inculcate politeness. A man should not attempt to force himself into the Museum unless dressed properly, say, as he would be going to church." Trustees James A. Garland and Samuel P. Avery expressed surprise that anybody "conducting himself in an orderly manner" should be refused admission to the Met, and all the city newspapers hurried to point out that the building occupied by the museum belonged to the city and that over two-thirds of its maintenance was paid by the city.

The day after the plumber's story appeared in print, every newspaper in New York City sent out reporters to test clothing-admission rules at the city's museums, restaurants, churches, and theaters. The *Herald* selected an African-American workingman, Edward McWilliams of 418 West Thirty-sixth Street, and accompanied him on a visit to the Met. McWilliams wore "a clean pair of blue overalls, a black coat and a new overcoat. On his head was a cap with a metal badge inscribed 'New York House Cleaning, Trucking and Carting Company, No. 5.' " D. W. Kellogg was making himself scarce, but Frank Marsden, his assistant at the turnstile, told McWilliams that he could not be admitted in overalls. Later Marsden explained to the *Herald* that he had excluded McWilliams because "he wore overalls, and then he was a colored man." When the *Herald* reporter demanded to know what difference McWilliams's skin color made, Marsden said, "Oh, I don't suppose it makes any." Eventually the overalls test case went all the way to General Luigi di Cesnola, director of the museum; he supported Marsden's action but denied Curator Arnold's earlier statement that women with shawls on their heads would also be excluded:

> Why, there are dozens of Italian women to be seen here every Sunday with shawls over their heads, and dirty shawls at that. You can see all sorts of ragamuffins here. We don't want to exclude the poor, but we reserve to ourselves the right to make any rules that we think necessary to protect the public.

And there the matter stopped. The Met went on record as concerned less with museum visitors looking at art than with the look of museum visitors. Protecting the public from the sight of workmen's clothing also entailed protecting art itself from being seen by workmen. Class and clothes, in this case, achieved distinct primacy over art. In a similar exclusionary move, middle-class-and-up public moralizers often bemoaned the desire of the lower classes to dress well instead of "saving" as they ought to, and complained about the social aspirations of "shopgirls" who dressed better than the well-to-do women they waited on. The moralizers' sword was double-edged:

they wanted to keep working-class Americans marked for exclusion, and when they were so marked they did indeed find themselves excluded.[22]

At the end of the century, when failure struck every level of society, big-city Americans became exceptionally sensitive readers of any sign that suggested decline and retrenchment. They noticed if a man wore last year's hat; noticed when his cuffs looked a little frayed; noticed that he had allowed his boutonnière to wilt. Furthermore, they read deteriorating objects and reduced spending not only as effects of bad times but also as their causes. For example, under the headline "How Panics Are Started," the *Herald* offered a parable of "Mr. Weed, a popular uptown tobacconist, sitting in his store reading his copy of the *Herald*," when in walks a regular customer who usually "buys his cigars by the box, and smokes none but the finest imported brands." The customer, looking gloomy and depressed, refuses to examine Mr. Weed's new Havanas, "a little ahead of anything in the market," and instead buys a cheap briar-root pipe and a dime package of tobacco. In the parable Mr. Weed decides that the cheap pipe "means something," so he cancels his own order for a "splendid nag"; in turn, the owner of the sale stable reads Weed's cancellation to mean that he had better forget about buying new clothes for himself.

The parable suggested that a national depression was not so much a complex economic reality as it was a chain of illusions, or perhaps a series of poor showings on the part of consumers who unaccountably stopped shopping for the best. The parable did not address the question of consumers' actual remaining resources.

Many whose resources had been erased nonetheless utterly refused to don the costume of retrenchment, always an ill fit for an American. In the single year of 1897, individual defalcations amounted to the equivalent of five hundred million dollars in today's money as desperate men battled to remain on the consumer level they had once reached, and perhaps even fought to continue rising. Bad times created a new social stratum: the disappearing class. Hundreds of men on both the high and middle rungs of the work ladder seized what they could from a club, or a bank, or a business, and vanished into Mexico or Canada. New illusions were put into play when men feigned fatal heart attacks in public and then hid out in the Adirondacks. Quite a few men were last seen toting large satchels—said to be stuffed with bonds, cash, and diamonds—onto ferryboats; their relatives later suggested that these absconders had perhaps fallen overboard in a fit of vertigo and drowned.[23]

There was no dearth of advice on how to spend when Americans' incomes were rising; after May of 1893, when incomes dwindled or disappeared, so did good advice on how to handle the situation. In 1893, Junius Henri Browne wrote in *Harper's Monthly* of the terror involved in maintaining an individual illusion of prosperity: "We live incontestably above our means, because our means seem insufficient, and we cannot adjust them to our ever-growing wants. Do we really know how to economize? When we attempt it, we certainly blunder. . . . We are raw novices, unable to learn." The *Herald*, in the national economic darkness of December 1893, fumbled with the notion of economizing but could frame it only as altered spending. One long and desperately cheerful article identified New York City shops that offered cheap knockoffs and imitations of desirable goods, while another suggested that consumers try ideological shopping: a man might find "enormous satisfaction in knowing that the ready made clothes which he wears are not the outcome of the sweating system" and a woman might adopt "opposition to the continual and needless wearing of furs" as her stance. The *Herald* also, and not without reluctance, suggested that such great motivators as fashion and desire might have to give way to other priorities: "While ministers preach and women weep, while turkeys are going the way of all that is good and railroad stocks are going the way of all that is bad, the shoes of the children are wearing out." Ministers such as the Rev. Madison Peters of the Bloomingdale Reformed Church at Broadway and Sixty-eighth Street did indeed preach about the "thousands ruined in their attempt to flourish in a style which is beyond their financial ability as well as their station in life," but Peters sideswiped his own point by suggesting that an American could actually be immobilized at a certain "station" in a hierarchical society. Next to no one cared to believe in that.[24]

Beneath persons struggling to buy shoes for the children lay those who were struggling, after May of 1893, to stay alive. Most curious of all responses to the plight of the big-city poor—more curious than the claim that help would hurt them—was the desire to see them, to take a tour of the never-risen and the fallen: to go slumming. Among the several voyeuristic activities called "slumming" in New York City was the visiting firemen's tour of saloons, amusements, and brothels. By the end of the century that tour was very old stuff, so standardized that versions of it appeared in such guidebooks as Ernest Ingersoll's *A Week in New York*. The twelve-page slumming tour that Ingersoll laid out was a venture into the past, dotted with the dives of long ago and the brothels of yesteryear. Always checking the clock, Ingersoll hurried his slummer past shops that were closed and away from foods that, if tasted, would assuredly make him ill; of most potential stops on his tour, Ingersoll judged that "a glimpse was enough."[25]

A newer style of slumming—the tour of poverty and ethnicity—took hold after 1893, and it was built on a new illusion: that an individual could do his bit for a social problem by *looking* at it. Slumming became a stylized performance of social helplessness, but important slummers, whether locals, outlanders, or foreigners, told themselves that they were in the business of examining social conditions. Consequently they required not a guidebook but a human expert who could show them more than the "front" of the city. In the matter of choosing a guide to the underworld, writers maintained the edge they had held for centuries. Journalist Richard Harding Davis guided novelist Paul Bourget; playwright Jacob Gordin guided novelist Henry James; and poet Richard Watson Gilder liked so much to be his own guide that he had to be persuaded to take a policeman with him "to save time."

A typical slumming tour began anywhere between nine and midnight at a police station, where slummers acquired an authoritative and protective police escort. The escort business was stylized and subtle, but it was a business, even if slummers rarely noted the moment at which money passed into the hand of the escort. First-time slummers were trundled past police-station photo galleries of criminals and then invited to examine assorted crimegoods: roulette tables, revolvers, burglars' tools, counterfeiters' molds, and always "a saw

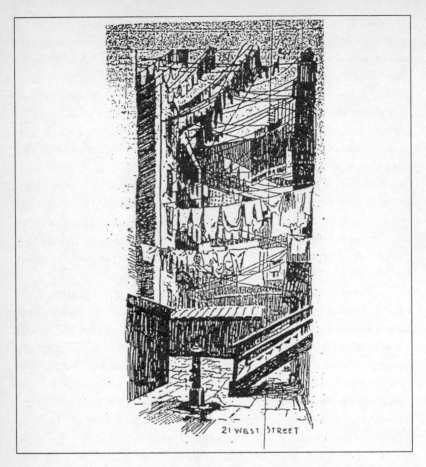

21 WEST STREET

with which a celebrated criminal had dismembered the corpse of his victim." The station-house visit put a dramatic and sinister frame around the sights to come—lower Manhattan's tenements, flop-houses, Chinese restaurants, Italian *trattorie,* and German beer halls. No slummer reported seeing any "criminal" activity or experiencing the slightest threat to his well-being, but each of them was encouraged to view his experience through the lens furnished by the police-station crime museum. From there, a standard slumming tour moved down into darkness, up into further darkness, and then out into light: down into dark cellars where slummers saw Chinese men smoking opium or gambling, and up into dark three-story lodging houses where slum-mers gaped at the filthy bare feet of sleeping men, and finally out into the light of public balls and beer halls and well-lit restaurants, for a conclusion said to be "cheerful" in tone.

Slumming combined social thrills with social avoidance; in effect it was all about looking—and only looking—at class levels lower than one's own. Rarely did a slummer actually meet or hear a word from a denizen of the world he was staring at. Slumming tours carefully pre-vented any verbal give-and-take by focusing closely on persons who were either comatose or sleeping. Not only did most slummers wear clothing that marked them as belonging to another world, but they also remained apart from the scene by hearing only voices that spoke authoritatively from their own familiar world. Paul Bourget, who slummed three times, did visit the Lower East Side in the com-pany of an "agitator by profession," a badly dressed and "disturbing personage" whom he called Bazarow, but Detective Clark, his police escort, continually interposed both his body and his contemptuous attitude between Bourget and the agitator. Police escorts and authori-tative guides, moreover, often insisted on matters that could not be seen at all: while Bourget was attempting to speak in their own lan-guage to a group of Italian immigrants, Detective Clark was simultane-ously busy trying to separate him from the group by informing him that these same Italians "sell their women to the Chinese over in the neighboring district. The law forbids yellow women to live in the United States. But John Chinaman has a great liking for white women, and he buys as many as he can with the money he earns or steals." When Bourget, the agitator, and the detective moved on to visit Chi-natown's stores, restaurants, and laundries, they spotted no purchased Italian women; by then, however, Bourget himself had been so drenched in the stream of race hatred that he lost his resistance and began to despise the Chinese for their obvious lack of "American" body size and their too-small feet.[26]

From another point of view, the notions that Detective Clark poured into Bourget's ear were very likely the thrilling stuff that he believed slumming white men were after, and about some of those white men he was right. In 1894, local labor leaders gave a banquet at

Clarendon Hall in honor of John Burns, celebrated British Labour MP, and were astonished when their honored guest suddenly disappeared. Burns claimed that he had slipped out because he found speeches and flattery boring, but in fact "he had arranged for a night much more to his own taste, spent in a trip through the lowest slums." On his tour, Burns made a show of comparing "social conditions" in New York to those in London, but his real interest lay in uncovering some imagined and potent combination of race, sex, and drugs. He stared in shock at a racially mixed group glimpsed in Chinatown, and "the question on which he could not get enough information was the cohabitation of white women with Chinamen. Mr. Burns was much surprised when told that most of the women in Chinatown could show certificates of marriage to their Chinese companions. He inquired minutely about the treatment of the women by their companions and the classes and nationalities from which they were drawn." Apparently his inquiries failed to satisfy the thrill-seeking side of Burns, for he was said to be "much disappointed that he had not been able to *see* more of New York's darker side."[27]

In 1894, entrepreneurs began to pick up on the commercial possibilities of race-based slumming. The *Herald* reported on "one of New York's Anglicized Chinamen who has a keen eye to business. He speaks English fluently and has the entrée to all the interesting mysteries of Chinatown. What he proposes to do is to make a business of conducting through the Chinese quarter select parties of sightseers" at nine dollars a head. "This charge is to entitle each visitor not only to a peep into the joss house and the Chinese theater, but it is also to cover the expense of a genuine Chinese supper with chop sticks and all the usual Celestial paraphernalia and delicacies."[28]

A thrill-seeker of a more mystifying order was the genteel poet Richard Watson Gilder, editor of the very genteel *Century* magazine. Gilder was a member of New York City's Tenement House Commission, and he used that official status not to study tenement conditions but to repeatedly make "midnight tours" of flophouses. Gilder admitted to puzzled reporters that the Tenement House Commission would make no recommendations concerning these establishments; he indicated that his "intense interest" arose from a "personal" determination to "observe how the poor of the city, the tenement dwellers, the tramps and all the homeless are sheltered at night." On a Christmas Eve slumming tour in 1896, the ambiguous Gilder visited eight lodging houses and closely followed the line that Detective Clark had drawn for Paul Bourget: Gilder saw the invisible when "in some places he guessed that liquors were smuggled in for the entertainment of lodgers, although no direct evidence of this was noticed." When reporters questioned him about what he had seen, he called his activities "a casual inspection," and further insisted that his observations were "too casual in nature to permit of an enumeration of evils." It is a rare nineteenth-century occasion when the word *casual* appears—as rare as an appearance of the adjective *safe*. When Gilder deployed *casual* twice in his remarks, he in effect stiff-armed the reporters and indicated that his slumming activities were closed to discussion. Then he slipped out of view behind officialese about the "best interests of the city" and lavish assurances that the commission would attach no blame to either individuals or corporations for "the evils of present conditions." Why this man, under the flimsy cover of an unrelated officialdom, returned night after night to stare at rows of men snoring and tossing in their sleep is a question whose answer no reporter could conjure out of Gilder's evasions.[29]

John Burns, MP, and Richard Watson Gilder, poet, appear to have been chasing special secret thrills when they went slumming; their half-concealed and unconfessable desires point up the troubled nature of the slumming tour and call into question its faith in the social value of staring. Some slummers were disturbed by their own activity. Paul Bourget, for example, framed his three slumming tours with many pages of statistics and sobering social thoughts so as not to "regret" what he himself called "very superficial experiences." Bourget nonetheless clung to his faith in staring when he insisted that he had gained "a more accurate view of conditions" even though the "details [he] was able to grasp were limited." Henry James, greatest of American novelists and most agonizingly precise of observers, refused to take the standard slumming tour: he judged "an inquiry into seaminess" to be futile, and he wanted no "gaping view of the policed underworld—unanimously pronounced an imposture, in general, at the best, and essentially less interesting than the exhibition of public manners." At several "characteristic evening resorts" on the East Side, however, James was unable to read "public manners." He stared and stared, and he finally admitted he could not read intelligence and ability from appearance alone. Perceptual failure, he decided, lay not in himself but in his subject:

> The answer had to be, for the moment, no doubt, that if there be such a state as that of misrepresenting your value and use, there is also the rarer condition of being so sunk beneath the level of appearance as not to be able to represent them at all. Appearance, in you, has thus not only no notes, no language, no authority, but is literally condemned to operate *as* the treacherous sum of your poverties.[30]

Thoughtful persons such as Paul Bourget and Henry James were discomfited by slumming, but Richard Harding Davis, the most famous, busy, and eager "guide" for important slummers, liked it so much that he made a second career of it. The now-forgotten Davis was, in the 1890s, a journalist and war correspondent of global fame and of a specific type: while Davis's rival Stephen Crane costumed himself as a bum in order to inhabit the flophouse scenes he wrote about, Davis used evening dress to set himself apart from flophouse scenes that he displayed to others. Although he puffed himself as a man equipped with special knowledge of the lower depths, Davis was neither liaison, communicator, nor bridge-crosser: he set himself up as an experienced and knowing protector of naive and innocent slummers, as a big Boy Scout who had won special merit badges visible only at night. When Davis led a slumming tour, he stuck firmly to his side of the social divide; under his care, theoretically, a slummer would experience no dangerous slippage in his identity.

Maintaining his knowing-but-uncorrupted persona took a toll on Davis. Paul Bourget noticed "extreme nervous tension, almost exhaustion" around Davis's mouth and eyes as he performed his standard tour opener, a "rendering of a coarse conversation, mimicked with a sort of genius." Davis was a Comstockian prude from whose good-boy persona nothing coarse could ever issue—especially not in the presence of admiring slummers. Under pretense of mimicking lower-class persons, however, Davis could simultaneously establish himself as an insider, tell a coarse story, avoid being thought coarse, and call attention to his "genius." His was a complicated act. He slung a tight web of invention around his slummers and enmeshed them in the invisible. Suddenly and without evidence, they could *see* a German saloonkeeper as just the sort of "crafty" person who "no doubt arranges those clandestine boxing matches which Davis has so accurately described in his Gallegher stories, tickets of admission to which cost a hundred, or even two hundred, dollars."[31]

The aura that Richard Harding Davis managed to drape over slumming tours did not transfer to his print efforts at same. In an 1893 *Harper's Monthly* narrative titled "The West and East Ends of London," Davis struggled to present himself as a man who moved with ease on both high and low social levels. He failed. Certain habits of reportorial honesty, in combination with habitual prurience, made hash of the suave persona he hoped to project, and he continually gave himself away. Davis reported that while he was "looking only to have a quiet smoke" at a high-class party in London's West End, he stumbled everywhere over couples engaged in "most embarrassing" sexual activities that he assured his audience were indescribable. Why a man seeking to smoke should be voyeuristically peeking into rooms on the third and fourth floors of a West End house he did not explain. The slumming-tour half of Davis's London essay ran into further trouble: in print, Davis was unable to frame his story with the coarse mimicry and self-referential patter that colored his real-life slumming tours. He could not, moreover, avoid shame.

Davis the writer chose to show himself intrepidly escorting a group of slummers on a Saturday-night visit to McCarthy's Lodging-House on Dorset Street in London's East End. The police inspector accompanying the group, however, was unfamiliar with the conditions desired by American night voyeurs; he must have thought that Davis actually wanted the group to see something more than Davis's own intrepid self.

At the lodging-house, the inspector woke all fifty men to exhibit them. They lay on strips of canvas naked to the waist, for it was a warm close night. They tossed and woke cursing and muttering, and then rested on their elbows, cowering before the officers and blinking at the light, or sat erect and glared at them defiantly, and hailed them with drunken bravado.

Davis was appalled; these men were not supposed to wake up and speak to—much less curse at—slummers who wished only to look at them. The horrible surprise cracked Davis's veneer of *savoir faire*, and he fumbled about for something to say. He hit upon a desperately ignorant remark that he considered "civil" under the circumstances: "The beds seem comfortable," he said to McCarthy.

In person, Davis had often succeeded at the pretense of knowing, but in print he was forced to admit that he knew nothing of the scenes he exhibited to others. "I confess," he wrote at the end of his essay, "to having in no way touched upon the East End of London deeply. I know and have seen just enough of it to know how little one can judge of it from the outside, and I feel I should make some apology for having touched on it at all to those men and women who are working there, and giving up their lives to its redemption." His apology had a short reach: it extended to a few unseen social workers, but not to the fifty men awakened for his entertainment.[32]

To most big-city Americans, *slumming* meant gaping and staring across the rigid boundary of race and the wavering boundaries of class. The term, however, continually shifted content as it moved between genders and across class levels. In the 1890s, public activities and amusements long closed to women began to open up slightly, but when something called slumming opened to them, it was nearly unrecognizable. "Real slumming" meant that men ogled safely, neither touching nor touched by anything they looked at, carefully maintaining their protected selves just as they did when they ogled Bouguereau's *Nymphs and Satyr* in the Hoffman House bar. Rewritten for women, slumming was hands-on and self-sacrificial, "hard and thankless toil" performed by "brave women" who battled against stiff resistance from fathers, husbands, and pastors. Stripped of opium dens, saloons, detectives, restaurant meals, and nighttime thrills, women's "slumming" turned out to be no more than a derogatory renaming of daylight social service, the kind of work that women dressed in the plainest of clothing had long performed.

While men's midnight rambles among the desperately poor went on unchanged, slumming as rewritten for women drew ridicule. "Slumming is not much in favor," sniffed the *Herald* in 1895, "yet some women prefer to humble the flesh by working among the very wretched." In an article ranking the appeal of various "Lenten diversions" for women, slumming came in just above roller skating but below bicycling and badminton. Women's "slumming" was further trivialized when the French chanteuse Yvette Guilbert sought publicity for a little day trip to Chinatown that she called "slumming." Guilbert, said to "love a lark," dressed for her star turn in a gown of silk and lace, a green velvet cloak, and a hat with a huge stuffed bird tilted on its brim. She drove to Pell Street in a "satin-lined perfumy brougham," had her fortune told, and listened to a recital of homicide narratives to which she responded with a lot of "oo-la-la."[33]

Slumming's reach was further broadened when the term was used to cover the old business of the rich entertaining themselves for the benefit of the poor. In December of 1894, according to an admiring *Herald* reporter,

> a group of young ladies and gentlemen hit upon a novel way of getting amusement and doing good at the same time on Christmas, not by going to the country, but by slumming in town. They are bound together to check all debasing influences of society and to elevate their own moral tone and that of the people who can be brought under their influence. They have collected about them a number of other young ladies and gentlemen of fashion, and gave a concert on Friday afternoon last at Sherry's. With the proceeds of the concert, which was rendered by professionals, they propose to give a dinner on Christmas day to the children of the poor attached to the mission of the Mariners' Temple, near Chatham Square, and the young ladies and the noblemen and gentry

attracted by the social distinction of the young ladies will serve the dinner themselves, acting as waitresses and waiters. Nothing quite so original has been attempted before.

The social pattern of the Mariners' Temple dinner was familiar: first the rich entertained themselves and their friends at a performance by paid professionals; then they used any remaining money to feed children removed from the context of their families. The mix of condescension and caution had a point: rich young people did not risk aiming charity directly at adults who might articulate the sort of unpleasant sentiments that small and hungry children were unlikely to utter to their benefactors. Like Mrs. Bradley Martin, they wanted to filter their social efforts and thereby protect themselves from the kind of disaster that struck Richard Harding Davis when fifty men at McCarthy's Lodging-House awoke and cursed him. Quite like male slummers who toured none but the comatose, the "young ladies and gentlemen of fashion" preferred to look only at persons not positioned to talk back.[34]

Benefactors in general felt more successful if the recipients of their charity could not speak. In *Park Row*, Alan Churchill recounts several tales of circulation-building charitable efforts dreamed up by William Randolph Hearst—efforts which, according to Churchill, "had a way of turning sour."

> One of his sob sisters later recalled how she was given a single can of ice cream to be doled out on a large Hearst-sponsored slum-kid jaunt to Coney Island: "All the way down there I was trembling to think what would happen when I dealt out that one miserable can of ice cream. When at last I placed a dab on each saucer, a little fellow in ragged knickerbockers got up and declared that the *Journal* was a fake and I thought there was going to be a riot. I took away the ice cream from a deaf and dumb kid who couldn't holler and gave it to the malcontent."[35]

Americans hungry for the thrill of staring at the hungry knew that children were their best resource. In 1893, Mr. and Mrs. F. W. Vanderbilt rented a hall and threw a Thanksgiving dinner for "the newsboys, bootblacks, messenger boys and other lads whose parents cannot afford turkey" at $1.25 a pound. Girls who were closed out of the employment available to boys were also closed out of the dinner, as were the boys' parents. The King's Daughters, an upper-middle-class philanthropic organization, carried out "to the letter" Mrs. Vanderbilt's ideas for the dinner. Leading citizens and clergymen paid tribute to Mrs. Vanderbilt, and Miss Amelia Tanner ensured that the boys observed "order and decorum." And of course there were those fully protected persons who chose only to look: a contingent of "well-known men and women" sat in a raised gallery and watched the boys consume "many tasty delicacies." After the meal, the lads "gave repeated cheers for Mr. and Mrs. Vanderbilt"—who did not attend. All pretense of seeing these children in their own environment—the pretense of slumming—was dropped here in favor of transporting selected urchins, *sans* possibly articulate older relatives, into the control of their benefactors and then exacting from them repeated displays of gratitude.[36]

At least the newsboys and bootblacks got something to eat. In a more horrific display of staring at the hungry, locals dropped by the toy departments of New York City's great department stores to look at poor children looking at toys. A *Herald* article on this version of slumming claimed that poor children learned early on that Santa "had no time" to bring gifts to them, so for them he left toys in the shops "on which they can feast their eyes and which to their minds means a great treat." One of those numerous "unwritten laws" that figured strongly in the lives of the poor mandated that poor children could linger in the toy department as long as they liked—provided they touched nothing. Children at the bottom of the class ladder were without live options: they could look and only look, and if they looked they were sure to be watched. "It is a lesson in itself," wrote the *Herald* reporter, "to see some of these tiny folk, poverty stamped on every line of their countenances, as well as in their ragged clothing, standing before toys, with their hands deep in their pockets or tightly clenched behind their backs." Some children waited for hours in hopes that a saleswoman trying "to tempt some rich customer" might pull the strings that made a toy cow moo and a toy elephant trumpet.

Then the children clutched each other in agonies of delicious fright, and the more dreadful the noise the more keenly did they enjoy it. Sometimes the little brothers and sisters taken with the elders, who have not yet learned that they cannot have everything, or even anything they want, will cry and stretch out their little hands for some toy that to their baby hearts seems most beautiful. The tact and patience shown by the child guardians to their little charges would teach many a lesson to those older and more blessed with this world's goods. The little ones are told that Santa Claus has just put the pretty things there to look at, and they must be glad that they can come and see them, and gradually the sobs cease and the little souls are comforted by this barren philosophy.[37]

The "hearts of onlookers" were "sometimes touched," it was said, by these Christmastime toy-department scenes, but not universally and not to any particular effect. The toy-department scenes were not about having and getting; their force would have been lost if some bleeding heart had actually put pretty toys into the hands of the sobbing children. Santa Claus mythology aside, these children were learning the discipline of looking, a harsh and demanding field of study that required its practitioners to combine great desire with great stoicism, and to believe unquestioningly in the beauty and power of goods. The discipline of looking demanded that individuals learn how to stare long and admiringly at goods that they were not to possess or even touch. The same discipline had been practiced, with varying degrees of success, by those who had read about, and then fought for a glimpse of, Cornelia Martin's wedding, and by those who had hoped for a peek at the jewels and flowers of the Bradley Martin ball. Poor children in toy departments were such successful practitioners—and professors—of the discipline of looking that well-off bystanders visited their toy-department *agora* to study, so to speak, at their feet.

Plenty of big-city Americans might have accepted the public rules of the toy-department scene while at the same time they prepared to violate those rules in private. To many men, looking was a mere preliminary to having, touching, and investigating what might be done with and to pretty young persons of a class beneath their own. Exactly like

Man Slumming, Man Looking sought the protection of his own kind: groups of disoccupied satyrs sat framed in the bow windows of Fifth Avenue clubs, or stood on street corners, or clustered at stage doors, there to appraise nymphs passing through their field of vision. They knew the stringent rules that governed public looking, and knew them so well that they arranged to do their touching far from public view. There, in private and among friends, few rules governed behavior.

NOTES

1. *New York Herald,* 30 January 1894.

2. *New York Herald,* 23 July 1893. French commentator Guy de Soissons thought that rich Americans especially did "not know how to distinguish a scion of a good, noble race, a real count, a real marquis, or a real duke. We say that one can feel a nobleman, but, if so, the instinct appears to be lacking in the case of the Americans" (86–87).

3. *Chicago Tribune,* 2 June 1893. In *America's Armories,* Robert Fogelson chronicles long-standing objections to any connection between armories and parks.

4. *Chicago Tribune,* 26 June 1893; *New York Times,* 11 May, 6, 11, 14 June 1893; *New York Herald,* 30, 31 January; 10, 18, 20, 27 February; 20, 27 March; 2, 8, 16, 20 April; 1 May 1893. In 1895, momentarily bored with reaming out Chicago, the *Herald* turned its sights on San Francisco, another perceived competitor, and publicized the views of William H. Chambliss, a former San Francisco social arbiter. According to Chambliss, in San Francisco society "prestige ranks according to wealth. The so-called fashionable set used to run a dishonest poker game, but that did not exclude them from society. Saloon keepers, gamblers and keepers of disorderly resorts are gladly welcomed into the set, if they have money. In New York you have rich vulgarians whose wealth is due to brains, but in San Francisco we have a complete mutton head, who simply stumbled across a lot of money. One self-styled leader of society claims to be a New York Knickerbocker but before he went West he ran a bootblack stand at Wall Street and Broadway. Many a broker whose boots he has blacked remembers him well" (15 March 1895).

5. *Chicago Tribune,* 5 June–7 June 1893.

6. *Chicago Tribune,* 10 June–13 June 1893.

7. *Chicago Tribune,* 14 June 1893.

8. All material on the Infanta Eulalie's visit is drawn from the following:

New York Times, 8, 11, 13, 15, 16 June 1893; *New York Herald*, 18, 19, 21, 22, 27, 28 May 1893; 1, 4, 5, 6, 7, 11, 13, 14, 17 June 1893. New York City news-papers that had made a business of scorning Chicago scorned Eulalie for scorning Chicago while they themselves continued to scorn Chicago. The *Chicago Tribune*, however, took the occasion of New York City's hottest June in two decades to point out that July and August would certainly be just as hot and the "fervent prayer of every New Yorker will be to get out of the city as soon as possible and stay away as long as he can. Happily there is a city of a refuge whose gates are open to them, where they can combine coolness and comfort with sight-seeing and instruction. That city is Chicago, where New Yorkers can live better and more cheaply than at home." New Yorkers, the *Tri-bune* generously asserted, were welcome "no matter how harshly they may have spoken of the Exposition in the past." Many went, though the *Herald,* at least, never admitted the fact. Manhattan was quiet and empty in August of 1893 only because, according to the *Herald,* the "country cousin is doing the great fair. Very likely he'll be sorry he didn't 'do' New York instead of being 'done' by Chicago. The fair is a fleeting show, while New York will be found at the old stand next season and many seasons yet to come, bigger, grander, more beautiful and more entertaining and instructive than ever she was." *Chicago Tribune*, 23 June 1893; *New York Herald*, 5 August 1893.

9. On Garret, see Elizabeth Drexel Lehr, 123–124. Another not unusual use of royalty was made by upper-class women deprived of both education and career: as an intellectual substitute, they memorized lists of British and European royalty and called it their "mental exercise."

10. *Chicago Tribune,* 2 June 1893; *New York Times,* 14 June 1893; *New York Herald,* 8 July 1893, 7 November 1893.

11. *New York Herald,* 1 December 1895, 29 December 1895. For a time in the 1880s, the Astors suggested that they were descended from the Royal Dukes of Astorga who had castles in Spain, but dropped the claim when the head of the Astorga family protested.

12. Iza Duffus Hardy, 98; Brydges, 116; De Rousiers, 340–341; Vay de Vaya, 383–384. The *Herald* usually did lower-class people in dialect, to make sure that class became available to the ear, and also to ensure that the dialect would override anything the person had to say that might otherwise be com-pelling.

13. *New York Herald,* 12 December 1895.

14. *New York Herald,* 31 December 1893, 21 January 1894, 10 March 1893, 15 April 1894; *New York Tribune,* 7 April 1897.

15. *New York Herald,* 14 December 1894, 28 September 1893.

16. *New York Herald,* 3 September 1894.

17. *New York Herald,* 6 December 1893.

18. *New York Herald,* 16 June 1893, 23–27 September 1894.

19. Francis, 202–209.

20. *New York Herald,* 11 February 1894. In *The World of Goods,* Mary Douglas further analyzes standard attitudes toward consumption: "Consuming at the same level as one's friends should not carry derogatory meaning. How else should one relate to the Joneses if not by keeping up with them? The popular literature on consumption is surprisingly supercilious about attempts to be equipped at the same level. Terms such as 'conspicuous consumption,' 'band-wagoning,' 'the Veblen effect,' and an aura of disapproval over keeping up with the Joneses puts the writers on the side of the excluders, against those who are trying not to be excluded" (125–126).

21. *New York Herald,* 25 August 1893. Reportedly Jane Addams, when lecturing to country girls who had come to the city to find work, invariably opened with "Ladies, your clothes are your address."

22. *New York Herald* and *New York Tribune,* 3–5 February 1897. In *The City Worker's World,* Mary Simkhovitch remarked, "If in general the working girl spends too much on her dress and cares too much for it, she is not thereby differentiated from other American girls" (130–131).

23. *New York Herald,* 29 October 1893, 3 December 1894. In 1880, Samuel Day discussed retrenchment in the context of an earlier American financial panic: "Has not Society established its canon against descending from one's high estate? Or, providing one daringly violates that stern ordinance, does not such a contumacious act notably imply loss of caste, loss of character, and as the sequel, loss of credit? Undoubtedly it does. Hence it is that merchants, bankers, traders, and others, upon whom serious pecuniary calamity falls, are ever loath to depart from their ordinary mode of life. The struggle is often severe—sometimes desperate. Sundry ingenious devices are employed to ward off the evil day until finally the unequal struggle can be maintained no longer" (II, 68).

24. Browne, 278; *New York Herald,* 3 December 1893; *New York Tribune,* 1 February 1897.

25. Ingersoll, 205–217.

26. Bourget, 190–198.

27. *New York Herald,* 6 and 9 December 1894.

28. *New York Herald,* 15 April 1894.

29. *New York Herald,* 19 March 1907. The scenes that slummers visited in the 1890s changed little until 1907, when health department regulations required clean sheets and pillowcases on the beds each day, fitting up of baths and washrooms, and ventilation appliances. The seven-cent canvas hammock

became illegal in 1907. Owners of the 105 lodging houses in Manhattan had to ask for permits to stay in business and could no longer defy regulations as they had for many years.

30. Bourget, 181; James, *American Scene,* 134, 193, 195.

31. Bourget, 193–202. In *The Reporter Who Would Be King: A Biography of Richard Harding Davis,* Arthur Lubow fully illuminates the ambiguous but not complex Davis. Stephen Crane's "An Experiment in Misery" (1898) furnishes a nice contrast to Davis's evening-dress slumming.

32. Davis, 279–292. In Davis's tours of New York City flophouses, he was accustomed to asking his tour group to glimpse the sleeping men "dimly," through "a vapor hardly pierced by an occasional lamp" (Bourget, 202).

33. *New York Herald,* 3 March 1895, 23 December 1894, 13 December 1895, 15 April 1894.

34. *New York Herald,* 3 March 1895.

35. Churchill, 80–81.

36. *New York Herald,* 3 December 1893.

37. *New York Herald,* 23 December 1894.

CHAPTER 5

Girls

The stage was slippery before I danced, and I complained. One of the gentlemen lifted me down while it was being fixed. I had shoulder straps holding up my dress, and one of the men asked what would happen if these straps were cut. They cut one of them, and—and—nothing did happen.

—Lottie Mortimer, *New York Telegram*, 9 January 1897

IN THE FINAL DECADE OF the nineteenth century, public lust for girls became socially legitimate. No girl was immune, and a man with specific preferences could publicly advise girls on how to improve themselves into more perfect objects of desire. On February 5, 1894, a twenty-six-year-old man dodged behind a pseudonym and fired off to the *New York Herald* a public plea to young girls: he wanted them to pay better attention to the nature of his sexual fantasies about them.

To the Editor of the Herald:

I was glancing over your new department, "The World of Womankind," two or three days ago, and enjoying it while riding up town in a Sixth Avenue elevated train. At Forty-second street three persons entered the car and sat down opposite me—a young man, a matron of middle age and the prettiest girl I have seen in a year.

By almost crossing my eyes I managed to gaze at her without appearing so to do, and indeed it was a treat. She was evidently a schoolgirl, about 17 years old, slenderly formed rather than otherwise. Her hair was brown, her eyes likewise, not over large, but bright and sparkling with vitality. She was graceful, was this little maiden, and her head unconsciously poised in curiously bird-like fashion, and she was also very well dressed; and I, a bachelor of twenty-six, sat there pretending to read the *Herald*, but feasting my eyes on this dainty picture.

Being a somewhat imaginative and impressionable young man, various possible and (I sigh to think of it) impossible things swirled across my mind. I wondered who she was and of what social station. I studied her brother's face so as to recollect it and obtain an introduction should I run up against him in a club.

These and thousands of other thoughts whirled across my mind, when, for the first time, she spoke, saying: "It was a lovely skate; but, oh! George, you just ought to have seen me fall!"

Crash! Went my mental pictures. Shivered were my hopes. The illusion was rent, not to be mended.

Not merely was the girl's grammar execrable, but her voice was the opposite of agreeable. Her tones were harsh, strident, irritating to any one used to refinement. Since this occurred I have noticed carefully the voices of girls in elevated trains and surface cars, and am surprised and disappointed to see how very many betray lack of culture. Nothing will so quickly show ill breeding in a woman as a harsh voice.

Pardon me for this hint, but I am sure other bachelors notice such things and are impressed by them. If the girls only knew how much men consider these "side issues" more care would be exercised I think. A beautiful girl, with a harsh voice, is like a basket of roses with a sunflower in the middle.

New York, Feb. 5, 1894. A. Bach.

A. Bach—or "Mr. Man," as Ring Lardner would come to call him— was searching for the perfect schoolgirl and thought he had spotted his prey: she was pretty, she was very young, she was well dressed, and she exhibited the slender girl's body that many men lusted for at the time. He needed only to know her social station in order to define her as suitable prey. She, on the other hand, was protected from him only by her sloppy diction and a voice that he found unpleasant. When she spoke, her voice communicated to him perhaps not so much bad grammar as her refusal to take her place in the sexual fantasies, the "impossible things," that he was busy building for her and for himself. The "matron" who accompanied her he judged to be no obstacle to his desires, and the brother, far from being a protector, he judged to be a possible route to the girl; were he a clubman, the brother would presumably be ready to hand over his little sister to a fellow clubman.

The idea that the girl was on display for assessment and potential annexation was so socially valid that Mr. Man could comfortably make public his designs on such girls. His letter was critical not of himself but of her: she should school herself, refine herself, and put on a better show so that his sexual fantasies and his future plans for her would not be shattered by the croak or squeal of a voice that he found unpleasant.

Mr. Man's desires were in no way unusual. Fascination with and lust for girls—even for girls far younger than the one on the Sixth Avenue elevated—were not merely American but global nineteenth-century phenomena. Lusting after children was publicly articulated in guidebooks purporting to warn against the phenomenon. Men with what were invariably called "proclivities" could use Howe & Hummel's *In Danger* to learn where to look for child prostitutes—Chatham Street, the Bowery, Park Row, Union Square, and Madison Square—and how to spot them. These girls might well be selling newspapers on the street: "Pretty girls they are, too, many of them, with large, lustrous eyes, long well-oiled hair, nice shoes upon their feet, short dresses disclosing evidences of graceful forms, ruddy complexions, and armed with many winsome little actions calculated to conciliate patronage. This life they pursue until they engage regularly in a life of shame." As was characteristic of Howe & Hummel, they compensated for having given out such information by censuring the social situation that had produced child prostitutes:

> It is a common sight to see children on the streets in all parts
> of the metropolis—boys and girls—aged from 5 to 15 years,

selling papers, shoplifting, stealing, and—worse. Have they parents? Who knows, who inquires, who cares? Some of them are very pretty girls, too. All the worse for them. The same causes which conspire to throw girls upon their own resources to gain a livelihood, operate with the brothers; but the latter are more fertile in means of accomplishing that end. Boys can black boots, sell papers, run errands, carry bundles, sweep out saloons, steal what is left around loose everywhere, but girls can only sell papers, flowers or themselves.[1]

Men with proclivities could view girls' bodies in displays of child pornography concealed behind green curtains at the rear of photograph shops; for public consumption the *New York Herald* parted the green curtains slightly when it began offering, in the late 1890s, full-page spreads of very young girls—children, really—photographed in grown-up poses of all sorts, many with their clothes apparently about to fall off. "Here is a group of some of the prettiest little people you ever saw" was the *Herald's* only comment on the purpose of these photos. A man who had a private gallery could indulge his proclivities amid nearly life-size representations of girls: shrewd French Salon painters like Adolphe-William Bouguereau produced for the American private-gallery market innumerable paintings of ever younger peasant girls with and without clothing. Men preferring music to art had their tastes addressed by popular songs such as the 1894 hit "Whose Little Girl Are You?"

Dreaming of love and of beauty
Dreaming of one sweet and fair
Dear little, sweet little darling
With blue eyes bright as the dew
Come, little one, now, and tell me,
Whose little girl are you?[2]

Men in groups made private arrangements to indulge their communal taste for young girls. When the satyr Stanford White spotted an ambitious, needy, and unprotected young girl, he signaled his friend the photographer James Breese to approach her and offer to produce at no cost the set of photographs she would need to launch a theater career. In contrast, an ambitious, needy, and unprotected girl like the sculptor Janet Scudder, whose body he did not wish to see but whose work he wished to use, White steered toward the tea parties of his wife, Bessie. In 1896, the seventeen-year-old Ruthie Dennis, later to become Ruth St. Denis, was occasionally performing in vaudeville while struggling to launch a career when she got a wire inviting her to dance at the Opera Club, a private group whose party at the Met began when the opera itself was over. At the Opera Club she met Stanford White. St. Denis described what happened after White pointed her out to James Breese.

> James Lawrence Breese was Stanford's running partner in the various exploits which made them so famous at this time— more than that, he was an excellent amateur photographer. Breese asked me to come to his studio on West 16th street. I went eagerly, with one object and only one in mind. I knew the value of beautiful photographs and I also knew that I could not possibly afford them at this time. . . . He asked me to come a second time, and on this occasion he stopped his restless pacing up and down the room and inquired in a charming, caressing voice if I would pose in the nude. He made it all very artistic and plausible. I had, he was sure, a beautiful body with long lines which he was anxious to capture. I was in a flutter of indecision for a moment or two, but vanity won out and I very chastely stepped out of my clothes. He took the pictures, I put on my clothes, and we did not meet again for many years.[3]

Breese—and other photographers such as Rudolf Eickemeyer with whom White had arrangements—then gave a set of photos to White to pass around among his friends or, depending on the nature of the photos, made a dozen copies of each photo so that each friend could have one for private viewing.[4]

Schoolgirls and their distinctive schoolgirl clothing appealed particularly to men who combined their proclivities with a certain class consciousness. Blanche Oelrichs, who described her girlhood self as "rather good-looking in a lusty manner," had been instructed by her mother that "a lady never looks into a club." In 1890s New York, however, she was aware that as she returned home from the Brearley School, she was being watched by men she was not supposed to look at: "All the gentlemen from the Union Club, anchored next to the Cathedral, knew the pretty school girls by sight, and the pretty school girls knew them." In 1894, in a new weekly feature titled "Through Club Windows," the *Herald* chose to defend clubmen against the "thoughtless and empty-headed persons who ridicule the habit that club members have of congregating in their club windows and contemplating the world as it passes by," and who called such clubmen "monkeys in a cage" and "cane-eaters." Truth was, according to the columnist, these "juries of captious critics scattered along Fifth Avenue at regular intervals" were performing the same service that Mr. Man sought to perform when he went public with his advice to schoolgirls: the clubmen were "working to improve the general appearance" of the women who passed through their purview. Men without the advantage of a bow window on Fifth Avenue took to congregating on West 125th Street between Seventh and Eighth Avenues, a lively and crowded area "built up with handsome houses and apartments." There crowds of "well-dressed mashers" lined the curb, ogled the "young girls who traversed the sidewalk," and offered "flippant and impudent remarks as to how they might improve their looks."[5]

The schoolgirl look—white platter collar, jacket, soft tie, round hat, and boot-grazing skirt—was the uniform of certified virginity, and its appeal to some men was so great that older prostitutes, according to Timothy Gilfoyle, "hoping to appease male fantasies for young sex partners," often disguised themselves to look like schoolgirls and appeared on the streets just as school let out. Evelyn Nesbit Thaw, sexual icon of the turn of the century, made shrewd use of the schoolgirl's appeal. Photographers posed Nesbit amid all the sexual freight of the time—youth, heavy hair, little bare feet, peasant blouses, kimonos, and animal-fur rugs—and she herself eventually added sexually spiced stories of showgirl life, drugged drinks, flagellation, and rape while unconscious. But in 1906, when she took the stand in defense of her husband, Harry Kendall Thaw, who had murdered the avid pursuer of young girls Stanford White, Nesbit wore schoolgirl costume. She thereby played a neat

double game: she asserted visible innocence for the benefit of the innocent segment of the public and she sought to arouse any man present
who found the schoolgirl look to be sexually exciting.[6]

A man could go public with his offer for a girl through the personals column of the *New York Herald*. Like Mr. Man in his letter to the
editor, such men invariably assumed an advisory or superior stance
and an arrogant tone in expressing their desires:

> *New York Herald*, 5 December 1897. A business man, 35,
> desires acquaintance healthy, sensible, pretty blond Protes-
> tant under 20; view, marriage; kindly describe appearance,
> age, height, weight; mercenaries, gum chewers or bicycling
> noodles not wanted.

Because, in the arena of urban exchange, payment is never made in
kind, Business Man revealed about himself none of the details of
appearance or state of health that he demanded of teenage girls. Moreover, since he clearly hoped to indulge his proclivities at small
expense—"mercenaries not wanted"—he aimed to sound businesslike
and even discriminating: his desired class level he encoded as opposition to gum-chewing and his distaste for the possibly too independent
athletic girl he slid under his scorn for "bicycling noodles."

Big-city men who, like Mr. Man, used public transportation to spot
likely women tended to use the personals to contact women whose
range of sentimental and sexual experiences was broader than that of a
seventeen-year-old schoolgirl. Men believed, or hoped, that women
were "noticing" them just as they were "noticing" women, even
though no woman ever advertised for further acquaintance with a man
she had seen on a streetcar.

> 11 December 1895. Broadway cable car, yesterday morning,
> ten fifteen—will lady wearing astrakhan jacket, who noticed
> gentleman who got off car at Court House, honor him with
> her acquaintance?
>
> 12 December 1895. Brooklyn Bridge, 3rd Avenue L Wednes-
> day, stylish hat, cape and dress—Your glances made such an

impression, I could not help but believe an acquaintance would be encouraged; if I am right let me hear from you.

24 December 1895. If magnificent stout lady who left Broadway car at Hilton, Hughes & Co's yesterday will walk down Broadway Thursday at 12, from 23rd street, the stout gentleman with full beard who sat opposite will try and meet her.

23 December 1896. Will lady leaving elevated car, 42d street, Monday evening, with companion, communicate with gentleman opposite who tapped her shoe?

Girls but a year or two older than the schoolgirl on the Sixth Avenue elevated, girls who knew that they had nothing to offer but their looks, their bodies, and their desire to please, used the personals to specify the nature of the exchange they sought, an asymmetrical exchange beneficial and pleasant to both parties.

4 December 1896. Refined miss, 18, plump, pretty face, good form, loving disposition, desires acquaintance gentleman of means; object matrimony. Girlish.

8 December 1896. A young lady, perfect figure, would meet elderly gentleman of means; object, matrimony. Brunette.

20 December 1896. A refined young woman of 19 wishes to meet well bred man who can appreciate and afford the luxury of a well groomed companion; object, matrimony.

A girl who offered herself in the personals was not an American Girl. That special category appeared after the Civil War and reached full bloom in the 1890s. Considerable cultural effort went toward insisting that the American Girl was an "ideal," but she was not; she was, in fact, an established reservation fenced in by specifics of national origin, race, class, marital status, and age. The American Girl was white, under twenty, unmarried, upper middle class or upper class, equipped with not an education but a list of "accomplishments," and not tied to any employment. She had been a girl since about the age of eight and, although she would remain some type of girl until she married, her status as an American Girl would weaken progressively after twenty. At thirty, if she had not married and was still living at home, she would become an elderly girl; if she left home to live on her own she was no longer a girl but a bachelorette.

As is not unusual in America, influential public speakers insisted on establishing a category—Women, the American Girl—and then began at once a furious effort at exclusion. The American Girl category was exclusionary along class, race, and nativist lines: it was never suggested that its white upper-class population held anything in common with immigrant girls, African-American girls, working girls, factory girls, girls in domestic service, or shopgirls. Because an American Girl had no sexual experience and earned no money, a child prostitute was therefore not a girl. Nor did an American Girl cross into the tiny and mysterious category of college girl, a "brainy" female human whose looks were believed not to be her top priority, whose designs and purposes in life could only be guessed at, whose class level was unclear, and who might well be past twenty. And last, there were showgirls, who might be as young as twelve, and who remained in the showgirl category as long as they remained onstage. A showgirl was probably not a Girl but both showgirls and American Girls were regularly called "butterflies": pretty and short-lived and not requiring consideration as serious human beings.[7]

Assessing the American Girl was practiced on an intercontinental if not even global scale, standard among natives and socially required also of visiting foreigners, social commentators, travel writers, and casual observers. In 1896, *Punch* cartoonist Harry Furniss—notably indulging in the end of the century's passion for defining national characteristics—wrote:

> When you are introduced to an Englishman, he invariably invites you to eat something. A Scotsman suggests your drinking something—urges upon you the claims of the Mountain Dew; a Frenchman wishes at once to show you something, the Bois de Boulogne or the Arc de Triomphe; a German desires you to smoke something; an Italian to buy something; and an

Australian to kill something, but an American wants an opinion right away. "Waal, sur, what do you think of our gre—e—
at country? What do you think of this wonderful city? What
do you think of the Amurrican gurl?" This latter is a question
which one is asked in the States morning, noon, and night.[8]

When pressed for a response to the question about the "Amurrican
gurl," Furniss's tactic was, he wrote, "simple: I hedged." Most did not
hedge but threw themselves, so to speak, onto the subject at hand.
Encouraged to see a category called the American Girl, they saw it. In
1892, H. Panmure Gordon, in *The Land of the Almighty Dollar*,
answered his obligation as a commentator on the American Girl, "the
most talked-of creature in the world," with close attention to both her
beauty and her extensive wardrobe. Gordon, however, could locate no
social function for the Girl in her present moment, so he reached into
her murky future: the American Girl, he announced, had "a consciousness that she can aspire to any position." In 1895, Paul Bourget—like Gordon apparently unaware that he had met and was
discussing white upper-class girls only—announced after lengthy
praise of the Girl's charm, grace, and daintiness that "the young
American girl is before all things else a whole little universe." She was
not a mysterious universe, however, for Bourget could "read upon her
charming face" everything about her that was worth reading. By 1898,
such remarks were so standard that James Muirhead, in his *Land of
Contrasts*, inferred that because every book on the United States contained a chapter devoted explicitly to her, the Girl must actually exist:
"The inevitableness of the record must have some solid ground of reason behind or below it. It indicates a vein of unusual significance, or at
the very least of unusual conspicuousness, in the phenomenon thus
treated of." After the turn of the century, foreign observers who had
met no American girls nonetheless pronounced them to be "very
pretty" and "vivacious"—the latter a much-used term that meant
extremely yet not annoyingly chatty. They further insisted on the
Girl's "absolute freedom" in American society, a freedom defined by
the Girl's "going about by herself, forming her own circle of friends,

her own interests and amusements, giving her own parties, and ordering and selecting her own toilettes."[9]

Paralleling the emergence of the American Girl was a growing insistence on a related idea: that men could not "understand" girls or women of any age, and indeed should not attempt to do so. Both men and women pressed this idea upon men; it did not carry the corollary that women could not understand men. Although it was obvious that no man could "understand" half of the American population, the idea was specifically converted into meaning that no individual man could understand any individual woman; any woman was and would forever remain to any man a mysterious and even alien being—and in that fact lay her ineffable "charm," a charm that she would lose were he to understand her. That final corollary strongly suggests that "the mystery of womanhood" was so trivial, so uninteresting, so possibly tiresome that it was better if the man never found it out, better that he never discovered just how empty she really was. The social effort to keep men and women emotionally and psychologically distant enlisted such powerful speakers as the social arbiter M. E. W. Sherwood, who announced that "the man does not live who can understand a woman." Newspapers and periodicals offered the same notion as the most current thinking on the subject of gender: "Not that any man was ever able to understand a woman, for he isn't. Ordinarily speaking, the more you know about anything the better you understand it, but when the particular thing is a woman then you don't. That is what floors you. Then you call it a psychological phenomenon and sit down." Such antipsychological psychology reinforced the American Girl as thing, and released men into a kind of paradise where they had only to assess the looks—and occasionally bend an ear to the charming chatter—of a decorative and useless group whose putative thoughts and desires were simply not available to the male mind.[10]

Most actual girls were shut out of the American Girl category, but they could be made to wish for it and to shop for imitations of its accouterments; those who had entered the category would be ejected when they aged, only then to discover that Being an American Girl had devoured the years during which persons of the other gender had got themselves educations and occupations.

Only when hard times came, as they did frequently between 1871 and 1907, did it become clear what kind of cultural product the American Girl was, and how serious were the effects of her focus on the brief span of her decorative possibilities and on shopping as personal enhancement. She had been allowed no other social effect; within her category she had been silenced, and though she had done a lot of "vivacious" talking in her time, she had not written. As the crash of 1893 deepened into a seven-year economic depression, many former Girls—those whose once-affluent fathers, brothers, and husbands had suffered business ruin—needed to earn and had no socially approved route to doing so. From 1893 onward, the *Herald* offered employment suggestions to them, at first suggesting such lines of work as "becoming a caterer in a way" or seeking to earn by "arranging floral trimmings for others" or becoming a secretary to a "society lady," a job represented as requiring only the abilities to imitate the employer's handwriting and keep many secrets. So many former Girls, however, tried putting pen to paper that in 1894 the *Herald* enlisted the popular poet Imogen Guiney to spread the news that "no woman can earn a livelihood at poetry" and to explain why she herself was seeking a paying job as a postmistress. In January of 1895, twenty-one months into the depression, the *Herald* addressed the subject directly, under the headline "Must Well Bred Women Starve?"

No account of the ways for making a living open to women is worthwhile which neglects work for the untrained—those whose need is sorest and most immediate. The most painful appeals of modern life come from women brought up in ease, who find themselves suddenly face to face with want, without the least ability or skill to serve the public. No one sees more of this distressful page than those who sit at editors' desks, for the woman or the man unskilled at everything else turns to pen or pencil, with the idea of doing something as good as the average print or illustration. The experiences of this sort last year are never to be forgotten. In the sudden failures of the panic it was heartrending to know or see women of ele-

gance, unused to a wish ungratified, in a week brought to
know their incomes cut off, their homes lost and their chance
of bread and roof depending on useless, ignorant, though
ever so willing fingers. What are you going to do with women
whose entire abilities are not worth one dollar commercially
to any living human being?[11]

The *Herald,* recipient of the handwritten effusions of many former
Girls and determined to discourage further such efforts, offered hun-
dred-dollar prizes for the best employment suggestion for hitherto
unemployed and untrained women. Early in the contest, entrants suggested work suitable to the way the Girl had presumably spent her life:
she might dress fashion dolls, string price tags, crochet borders for
underwear, design neckties, and do art needlework. Then the suggestions began to move her out of the home. The former Girl, it was
suggested, might deliver services to those who had not yet been financially ruined: "cosmetic artistry" demonstrations, electro-therapeutics,
gymnastics training, mechanical massage, fragrant baths, and facial
damasking. She might become a professional shopper. By the end of
1895, mundane and productive activities for which the Girl had no
training began to appear among the contest entries: she was told to
attempt mushroom cultivation in the basement, make aprons, test and
sell cooking utensils, and raise poultry behind the house. Never once
did a contest entrant suggest that the former American Girl might seek
a paying position, cross over into the socially distinct category of
Working Girl, and type for eleven dollars a week while the man seated
at the next typewriter earned fourteen.[12]

Under extreme social stress, the American Girl nonetheless grew
bigger and whiter in the work of such illustrators as Charles Dana
Gibson and Howard Chandler Christy. Like Mr. Man, who conducted
a "careful" study of his schoolgirl subjects, both Gibson and Christy
set up as authorities on the Girl. The *Herald* classified Howard Chandler Christy as a "careful student of sociology who probably knows all
that man can know about that delightful variety of womanhood which
has baffled, bewildered and bedazzled two hemispheres—the American Girl." Neither man, however, was expected to display any under-

standing of his subject; instead, both offered the Girl as racial category. While the Anglo-Saxon whiteness of Gibson's girls was encoded as "patrician," Christy's view of the American Girl was openly racist. Christy announced that the Girl "has in her veins the vigor of Northern blood," and had "been led to seek her best development through hygiene and, beginning with no higher desire than health, she has discovered for herself that the fairy Health brings in her hands the priceless gift of beauty. Not a prude, she yet enforces the respect that is the tribute to her pure womanhood and enjoys the freedom that comes from fearless innocence rather than enforced ignorance." Christy's Nordic model, surrounded by a language of *blood* and *health* and *purity*, went on to take her place among the images of twentieth-century fascism. In America, at the turn of the century, she was a big healthy girl, still interested only in herself, covered from chin to wrist to toe in hygienic white shirtwaist and practical black skirt; she knew about sex but was having none of it. Richard Harding Davis, the sexually timid Boy Scout and celebrity racist guide to the slums, served as model for the male companion of the big and sexually unthreatening American Girl.[13]

Other men liked girls less armored in their clothing, needy girls, girls who could be persuaded to perform and to undress for them. These men made no public statements about girls; they listened to others talk about girls while they silently contemplated action. They were, however, but one group of contestants in the cultural tug-of-war over how best to consume women between the ages of fourteen and twenty. Other contestants were sweatshop operators, store owners looking for cheap labor, rich women in need of domestic help, theatrical agents and "entertainment caterers" looking for young bodies and innocent faces, middle-class reformers trying to tuck young women into Girls' Clubs where they could be "refined," and lawyers in the breach-of-promise business seeking young women to swear out lucrative affidavits against rich old men. Everyone, it seemed, wanted girls' services, wanted to get everything they offered before they turned twenty, and wanted all of it cheap. Girlhood was a narrow corridor with a fast-slamming door: Cornelia Martin, who was sixteen at the time of her marriage to the Earl of Craven, was called "a little girl" in the newspapers, but Consuelo Vanderbilt was judged to be "not so young by years as she is made out to be. She is very nearly eighteen years old."

The brevity of girlhood was asserted as a national physiological fact; girlhood was a stage followed immediately by the toothless hideousness of old age with nothing between, no middle age, to separate freshness from decay. "The face of an American girl, as well as the voice," wrote Harry Furniss, "is often that of a child. But suddenly, ten years before the time, and in one season, the bubble bursts, the baby face collapses, just as if you pricked it with a pin, and she is left sans teeth, sans eyes, sans beauty, sans everything." In his twin preoccupations with young girls and ugliness, Harry Furniss was a representative nineteenth-century man. No sensible such man would choose to restrain himself until the girl of his choice underwent the early and inevitable collapse. The girl's moment was always exactly *now*.[14]

Over an 1893 story about a society *tableaux vivants* show that had significant erotic content, the *New York Herald* headlined:

YOU MIGHT LOOK BUT NOT TOUCH.

To look at girls—in Gibson and Christy illustrations, in photographs, in Salon art, on the street, on the elevated, onstage—was socially valid, expected, even demanded. To touch them was, in a society both terrified of and fascinated by touching, another matter. *Fin de siècle* America had no available functional public language for the act of touching and no speakable names for body parts that might be touched. Events that involved touching vanished into crevices of inarticulation; public discussion and investigation of behavior came to abrupt halts when touching—called "the indescribable"—appeared. The public censors who had robbed America of a publicly usable language for touching had not thereby stopped people from touching each other; it is likely that, by dropping the veil of social silence over touching, the censorious had actually protected those who were dedicated to brushing against, touching, and groping. If no one could publicly name the activities, no one could put a crimp in them.

In December of 1896, there occurred a private social event—always thereafter known as the Seeley dinner—that enfolded into one package the late nineteenth century's hot cultural struggles over men's established practices of entertaining themselves in groups out of view of their families, men's desire to examine and touch the bodies of young women, extreme class and gender divisions, the enforced discipline of looking, and the fear of touching. Draped over all of it were familiar Orientalist fantasies, Salon-art images of satyrs surrounded by multiple nymphs, and the ever-expanding cultural uses of the two multilayered popular fictions that nearly everyone had read or heard about: *The Thousand and One Nights* and George Du Maurier's 1894 best-seller, *Trilby*. As a late-nineteenth-century event, the Seeley dinner encapsulated all the notable hungers of the time, the special difficulties of satisfying those hungers, and the absence of a language available for speaking of them.[15]

On the evening of Saturday, December 19, 1896, Herbert Barnum Seeley threw a stag dinner for his soon-to-be-married brother Clinton Barnum Seeley, at Sherry's restaurant, trusted venue for major society events. The Seeley brothers were well-to-do nephews of the late P. T. Barnum. Both were members of the Larchmont Yacht Club, as were the twenty invited male guests, half of whom were married. Herbert Seeley ordered up a thirteen-course dinner, and then spread the capacious umbrella of *art* over his entertainment arrangements when he enlisted Theodore Rich, a member of the Art Committee of the Larchmont Yacht Club, to make sure that "everything should be done in first-class style." But when Seeley and Rich set about explaining to various Manhattan theatrical agents and "entertainment caterers" the exact combination of storytelling, singing, and dancing that they wanted to buy, some agents said that they "did not deal in that sort of thing"—a sort of thing whose nature emerged gradually over the course of the next six weeks—and others demanded, for such special arrangements, a higher price than Herbert Seeley cared to pay. Later and in public Seeley refused to define the "sort of thing" he had tried to buy from men who had "that sort of thing" to sell. He would say only that he had wanted to book "not a Sunday School" but something "hot"—something like Lottie Gilson, the "Little Magnet," singing "You're Not the Only Pebble on the Beach." That popular 1896 song, billed as "the greatest laughing success in years," must have signaled to Herbert Seeley that Lottie Gilson understood both his upcoming stag party and her place in a stag world.

> *I was listening to a talk between two men, the other day*
> *The conversation ran on married life;*
> *And I was interested as I heard one of them say*
> *He thought that every man should have a wife.*
> *For he said, "My friend, I'm married, and I'm as happy as can be;*
> *But don't let it go farther, I beseech!*
> *I haven't seen my darling wife in years, 'twixt you and me,*
> *And there are others like me on the beach!"*
>
> CHORUS
> *There are a lot of others on the beach!*
> *And you can take advice from what I preach:*
> *When on married life you start,*
> *Take a "tip" and live apart—*
> *For there are lots of other pebbles on the beach.*[16]

Eventually Seeley and Rich stumbled on a man ready to supply what they wanted: James H. Phipps of Phipps and Alpuente, theatrical agents, 21 East Twentieth Street. Phipps agreed to send potentially congenial performers to "audition" at Herbert Seeley's bachelor apartment; there Seeley might also discover whether they understood, without his having to articulate it, what he wanted them to do at the dinner.

Whether or not Herbert Seeley's entertainment requirements went beyond the usual, there was nothing unusual about the dinner itself. Private men-only dinners, for any occasion or for no occasion, had been a feature of big-city life since at least 1870; in the 1890s, as women began to attend more and more public amusements that had previously been closed to them, men answered by creating more and more private entertainments for themselves. By 1893, men were choosing not only to seek amusement away from the other gender but also to eat away from it. The University Epicurean Club, an aggregation of seventy-five millionaire merchants, men-about-town, and politicians, all led by Moses R. Phillips, bought fifteen thousand dollars' worth of cooking equipment but no forks, and installed the equipment and themselves on the upper floors of a building on University Place near Union Square. The Epicurean Club's private space allowed them to satisfy a particular appetite: each man gripped a tankard of ale in one hand, seized a broiled steak with the other, dipped it into a vat of "epicurean sauce," and went to gnawing on it. Exclusionary club rules contained but one exception: "No woman is allowed within the hallowed portals except Rosa, the cook, who is the instrument in Phillips' hands to carry out his orders for the gratification of the members. She is a Virginia negress and with her wonderful 'old mammy' ideas of kitchen practice is an invaluable aid to the enthusiastic efforts of Phillips. She is an enthusiast herself, and the Epicureans look upon her as a jewel." The *Herald* found it "interesting to note that the wives of the members have shown a determined opposition to the club's existence. They say that its pleasures and its attractions are denied to them, and that since it was formed their husbands have shown too much of a tendency to eat there, and there only. But their opposition is vain." Clubs devoted to gross feeding became the rage of the moment: in February of 1894 the Gridiron Club invited members of the Turtle Bay Beefsteak Club of New York to travel to Wormley's Hotel in Washington, there to teach them and their guests—cabinet officers, senators, and congressmen—how men ate in New York. The Turtle Bay boys obliged with a theatrical display: they cooked enormous steaks in the presence of the guests, "served them without the accompaniment of knife, fork or plates," and awarded an engraved silver medal to the man who gobbled down the most meat.

By 1896, the nature of men's private entertainments was changing radically. Cigars and an after-dinner jokester were out; professional performers drawn from the legitimate theater, from vaudeville, and from a nether world of entertainment were in. These dinner entertainments were a function of the bachelor life, a life that a man, whether married or unmarried, indulged in with male companions all known to him and all of exactly his own class level, out of view of women of his own class, without their knowledge of any of its specifics, and without reference to their wishes. The bachelor world became a known world whose exact arrangements were nonetheless shrouded in silence. Bachelors themselves employed a generalized language of *sprees* and *gaiety* and *high jinks* when they arranged for, and notified each other of, "interesting performances"—not only elaborate dinners and vaudeville but also human sex shows, animals copulating, young African-American boys hired to punch each other into blood-spattered unconsciousness, and caged circus animals that men could tease into anger.[17] Certain agencies, according to the *Journal,* "did little else than arrange affairs of this kind"; they listed performers that the "general public knew nothing about," but who had "found that they can make more money by appearing at two or three club and private shows during the week than by seeking regular engagements at regular theaters." The bachelor life had but two rules: wear evening dress and keep mum. The Seeley-dinner affair violated both of those rules and thereby brought the bachelor world of the late 1890s into public view for the very first time.[18]

Near 11 P.M. on the evening of December 19, 1896, William S. Moore of 207 West Fortieth Street arrived, in an excited state, at the Nineteenth Precinct station in the Tenderloin, where he told Precinct Captain George S. Chapman that something "indecent" was going on in a private dining room at Sherry's restaurant. Little Egypt, according to Moore, had been booked to dance naked at a private event. Moore also told the captain that theatrical agent James Phipps had offered fifteen dollars—equivalent today to about $250—to Moore's own daughter Annabelle, a beautiful and innocent eighteen-year-old girl, if she would dance at the dinner and "arrange it so that her tights would fall off while she was dancing."[19] Annabelle had felt insulted, and when she told Moore about the offer he felt insulted too. Only later would it emerge that Moore himself had dispatched Annabelle to Phipps's office and that Annabelle herself would have felt less insulted had Phipps offered her twenty dollars. Moore indicated to Captain Chapman that only moral outrage had driven him to visit Sherry's on his own to put a stop to the show. At Sherry's, however, Manager

Eugene Flauraud had dismissed Moore; an event in a private room at Sherry's, said Flauraud, was none of Moore's business and, for that matter, none of Flauraud's either. Moore, for the moment, wore the cloak of guardian of the public morals, a garment available to anyone wishing to assume it at the time; several principals in the Seeley story would eventually don it.

Around midnight, Captain Chapman, wearing "citizen's clothes" and accompanied by Detectives Walter and Caddell, set out to look into Moore's complaint. Chapman had no warrant and would later argue that he needed none; because Sherry's had a liquor license, he could inspect the premises at any time. Although the door on Sherry's Thirty-seventh Street side was supposed to be a fire door, Chapman and the detectives found it locked; they heard sounds of clapping and shouting emanating from somewhere within. Through a window in the door, Chapman showed his shield to Assistant Manager Wilson, whereupon Wilson turned and raced up a flight of stairs. The clapping and shouting ceased; Chapman and his men waited. When Wilson returned to usher out an unidentified man and woman, Detective Caddell rushed the door and thrust his foot in it; the three men then entered the building and went upstairs unaccompanied, Wilson having again vanished. At the top of the stairs, Captain Chapman asked directions of a waiter, who pointed him toward not the banquet room but the dressing room. There he stumbled in on either six or eight women in states of undress, a maid, a female child, and four fully clothed men, two of whom were busy attaching small gifts to a Christmas tree. What Captain Chapman saw before the girls dashed behind the portières and who spoke angrily or flippantly to whom remained matters of contention to the end. One of the men in the room, Horatio Harper of Harper & Brothers, threw a punch at Captain Chapman but was too drunk to land it. Theodore Rich of the Larchmont Yacht Club Art Committee tried to reassure the captain by inviting him to examine a Trilby costume worn by Minnie Renwood, one of the hired performers. Captain Chapman, however, was that rare American who had never read *Trilby*, and he could make nothing of the signs pinned here and there on Renwood's person. Rich made a second attempt to dispose of Chapman by asking him to step out into the hall and examine "a menu that would tell what a lot of prominent men were at the dinner." Chapman, however, brushed Rich aside and entered the banquet room itself. There he assumed the guardian-of-the-public-morals role and delivered to the dinner guests a speech critical not of them but of the performers: "Gentlemen, the woman who would so degrade her sex as to dance naked before a party of men is a direct insult to your wives, your mothers, and your sisters." The men received his remarks politely and indeed pressed Chapman to join them, see the performance, smoke a cigar, and have a drink. The captain, however, announced that he was teetotal and departed.

By Monday, January 21, Saturday's polite tone had dissipated, the Nineteenth Precinct station was thronged with reporters, and newspaper stories characterized Chapman's actions as an unjustifiable raid. Many of those involved in the affair expressed indignation and outrage. Manager Flauraud asserted that "there had never been a breath of scandal against the place before, and that Mr. Sherry was very particular about the house and everything that went on there." Louis Sherry himself said that he "catered to the very best people in New York and to no others. The gentlemen at that dinner were of the very highest standing." Though it was already apparent that Assistant Manager Wilson had been assigned to lock and guard Sherry's fire door during the Seeley entertainment, Sherry nonetheless threatened to make out a complaint against Chapman and retained a lawyer to bring suit against the city. Cora Routt, who may have been barebreasted when Chapman entered the dressing room, also hired a lawyer. Phipps, the entertainment caterer, suggested that Chapman's actions were part of a "plot to injure his agency, launched by some rejected performer," and denied that any such person as Little Egypt had performed at the dinner. The *Herald* claimed to have made its own "careful investigation" of the matter and could see "no reason for such an attack upon one of New York's most exclusive resorts." The dinner guests, it was predicted, would all be filing affidavits and lodging formal complaints against Captain Chapman, who would himself most likely be banished to "Goatville." Meanwhile, in the first sign that something was not quite as it was being made out, Herbert Barnum Seeley fled to Boston and could not be reached.[20]

In the days following, not one of the guests at the Seeley dinner appeared at police headquarters on Mulberry Street to make a complaint against Captain Chapman. Once sobered up, the men had apparently spotted certain weaknesses in their publicly outraged position; from that point onward they labored to tug back over the dinner the rapidly slipping veil of silence. The bumbling Herbert Barnum Seeley himself broke that silence when he mailed to police headquarters a statement in which "he endeavored to show that while something out of the ordinary was expected in the way of dancing specialties, it was positively not stipulated that any performer should appear in the nude." The Seeley boys became figures of fun: on December 23, a group of brokers on the Consolidated Exchange "beguiled the dull hours of the afternoon by dancing the coochee-coochee around Clinton Barnum Seeley, whose naturally florid face was crimsoned by the attention that was directed to him."

Meanwhile the newspapers dug around the edges of the momentarily dormant story. The *Journal* detailed an earlier party adventure of Seeley's:

> Under the pretence of taking in new furniture Seeley smuggled a number of wild beasts into his apartment and enjoyed himself hugely by placing an unsuspecting chappie near a concealed lion or tiger and then making the animal roar. It was too demnition funny for anything until the guests took a hand at making the animals roar and a young lion escaped from its cage into the apartment. Then the guests roared. So did Seeley. They hustled into the sleeping room, climbed up on the mantelpiece, cut out into the halls and bawled for help until the keepers of the animals came to their rescue and caged the lion.

The *Herald* unearthed Herbert Seeley's previous involvements in turf scandals, and further discovered that Annabelle Moore was actually Annabelle Whitford, daughter of a Chicago telegrapher. It appeared that William S. Moore, once a family friend of the Whitfords, had lured Annabelle and her mother to New York, where he put Annabelle on the vaudeville stage at the age of twelve. Although Moore—henceforth denominated Annabelle's "alleged father"—claimed to be himself a theatrical agent, he had but one remaining client; he wept when made to admit that he had been living on Annabelle's earnings for the past two years.

By December 24, not one outraged dinner guest had appeared at headquarters to file an affidavit, although both Annabelle Whitford and Little Egypt herself had done so. Chief of Police Peter Conlin, by then impatient, suspicious, and exceedingly well informed, publicly urged all the Seeley dinner guests and Louis Sherry to drop by police headquarters to file their affidavits. He showered them with three letters apiece. To a man, they refused to appear as complainants, and some of them hired lawyers. On December 31, the day after his brother's wedding, Herbert Seeley, against the advice of his lawyer, appeared at headquarters to make an affidavit about the dinner he had hosted; it was rumored that in the affidavit the craven Seeley implicated several of his guests in lurid doings. Headquarters had released not a word of Annabelle's and Little Egypt's affidavits, but did release a part of Seeley's in which he characterized his guests as a group of elderly and innocent fellows—"all the way up to 55." The dinner entertainment, according to Seeley, had been so dull that the men had barely glanced at it; he admitted to the presence of Little Egypt, but said she had "just wiggled a bit, this way and that, and it was uninteresting, as they all had seen that sort of thing before. There were cries of 'Take her off!' and so on." Captain Chapman's entrance, Seeley complained, "threw a great deal of a damper" on an already quiet affair, "and the spirits of the company could not be restored." The frightened Seeley, however, chose also to offer perhaps the only information capable of shocking his dinner guests and men like them: he listed the names, addresses, business connections, and marital status of each of the guests at the dinner.[21]

The club world reeled at such a breach of tacit agreements and silent loyalties. The *New York Tribune* reeled along with its clubman readership: "The confession of Mr. Seeley caused a great deal of talk in the clubs last evening, and the comment of clubmen was that a statement more needlessly full of details and more certain to bring ridicule on the author and upon his friends never was heard of in this city. It was predicted that every one of the men named as guests at the dinner would be called as a witness at the trial of Captain Chapman and be subjected to questioning and cross-questioning about the performances that he was invited by Mr. Seeley to witness. Mr. Seeley, it was declared, would have himself to thank for all the trouble, and his confession would be a warning and a nightmare for a long time to come."

At this point in the affair, the *New York Times* editorial page seized the position of guardian of the public morals. After carefully listing the legalities of the case, the *Times* leveled its shot at Sherry's: "A good many people who are not indifferent to the moral decadence of the community would like to know whether it is customary with houses of entertainment at which fashionable dances and receptions are held to cater also to the repulsive trade of persons who want indecent dances with their dinner." In the next column, the *Times* reinforced its moral-guardian position with a letter from a grateful subscriber who pronounced the *Times* "a clean, wholesome family paper, bound to be appreciated by the many thousands who have become ashamed to place some of our great newspapers in the hands of their children." Across the country, the Seeley dinner occupied the front pages of newspapers that used it to take potshots at New York's social rich. The *St. Louis Post-Dispatch*, below the headline "Swell Set Naughtiness," announced that the Seeley dinner revealed "the utter moral rottenness of the so-called 'smart set' of Gotham, as shown by the carryings on at one of the swellest resorts of aristocratic Fifth Avenue, a place frequented by the grand dames of New York and supposed to be altogether above reproach. The trial—if the most powerful society influence in New York does not succeed in staving it off—will reveal more of the innate nastiness and hypocrisy of New York society than was ever before laid bare."

The *Post-Dispatch* reporter hunted up Little Egypt, whose name was Asheya Waba, found her living under the name of Mrs. Harper in an apartment house on Seventh Avenue, and learned from her that she had been paid one hundred dollars to perform two dances. The first of these was to be executed atop a table, and she was to wear "a little velvet jacket that would simply cover her shoulders," loose gauze trousers, black stockings, and slippers. During this dance, she confided, the Seeley dinner guests began to shout to her. "They want me to dance in the altogether," she said. "They call me Little Altogether

Egypt." Just as she, who "loved to dance in the altogether," was about
to comply, however, Phipps, the entertainment caterer, burst into the
room to shout that there was trouble. Some of the men rushed her out
of the banquet room and during the next two hours shifted her from
room to room in the nether reaches of Sherry's upper floor, plying her
with champagne until they thought it was safe to send her home. The
day after the dinner, according to Little Egypt, she had a visitor—
undoubtedly Herbert Barnum Seeley.

> A man came to my house. He was at the dinner. I saw him
> there. I was surprised he knew my address. He say to me,
> "Little Egypt, you must go away. You must go out of the state
> at once. There is trouble." He pleaded with me. He wring his
> hands like this. Oh, he begged so hard. I told him I could not
> go. I say, I have other engagements, I must keep them. He say
> "Little Egypt, you must go. I pay you money." He offered me
> a thousand. He showed me the money. I did not take it.

Police Chief Conlin tired of waiting for the dinner guests to file
complaints; he was, in fact, certain that public airing of the case
would clear the police department. Urged on by Police Commissioner
Theodore Roosevelt's demand for a "thorough investigation," Conlin
himself preferred charges of "conduct unbecoming an officer" against
Captain Chapman and ordered Complaint Clerk Petersen to make out
forty-four subpoenas.

When the Chapman hearing opened on January 7, 1897, nearly
four hundred persons jammed the stuffy hearing room at police headquarters: Seeley diners under subpoena, every police headquarters
official who could steal the time to be there, squads of lawyers and
reporters, and, according to the *Telegram,* throngs of "dandies, degen-
erates and patrons of the ballet." The men in charge tacitly agreed that
the hearing, like the dinner, would be a men-only event: they ejected
two women reporters from the *Journal,* and the sole—and heavily
veiled—woman spectator in the room chose to depart also. Seeley's
dinner guests, denominated the "chappies" by the police, were resitu-
ated in the adjacent School of Instruction room, where they sat

throughout the hearing. Each day at noon, the "resplendent chappies" were made to file through the packed hearing room on their way to lunch and to step outside to the jeers of the assembled hundreds unable to gain entrance to the hearing. Only once did one of them speak. On the first day of the hearing, a "chappie" asked, as he passed Chapman's lawyer William Hart on his way through the room, "Can't we fellows stay and see the sport?" Hart retorted sharply, "No sir, certainly not." Thereafter the chappies used their lunchtime file-through to cast reproachful glances in the direction of Herbert Barnum Seeley, their dinner host.

The *Herald* reported whispers that Little Egypt might be "on hand ready to do a turn in gossamer for the enlightenment of the commissioners," and indeed "expectation in the hearing room rose to the point of delirium when Colonel William F. Howe announced in his deepest bass, reinforced by diamond flashes, that he was appearing for Little Egypt." Howe understood perfectly, just as did Mrs. Bradley Martin, how late-nineteenth-century lighting worked with diamonds to create personal spectacle.[22] Howe's jewelry was, however, the day's only entertainment; opening statements and arguments were so dull that even Commissioner Roosevelt yawned repeatedly. Nonetheless, expectation of a show did not diminish among reporters, many of whom had managed to get a look at Little Egypt's affidavit. The *St. Louis Post-Dispatch* reporter, lifting freely from the *Journal*'s coverage, saw the hearing's first day as "only a preface to what promises to be as joyous a story as was ever penned by Boccaccio," and as a "spicy opening, ending in a lot of veiled promises."

After its opening day, in fact, the hearing ceased to focus on Captain Chapman's procedures at Sherry's on December 19; it became instead an extended inquiry into who had arranged for whom to see what, accompanied by investigation into the transparency of various fabrics and minute questioning about which performers wore tights. The testimony was confused always by the absence of publicly usable language and by men's inability to name body parts that might have been visible to the diners. The *Sun* characterized the lawyers' questions to the witnesses as "filthy," and the *Tribune* found the witnesses' responses to be "pure filth." The *Telegram*, meanwhile, classified both the dinner and the trial under "amusements," and the *Herald* took a balanced view and called the whole affair "amusing, outside of its filth." Under pressure from the facts, an event that had begun in the headlines as "The Chapman Hearing" became "The Seeley Dinner Trial."

On January 8, the men-only show began when testimony alluded repeatedly to "suggestive" signs pinned to Minnie Renwood's Trilby costume, and to what Herbert Seeley had meant when he requested entertainment with some "ginger" in it, and to the nature of the "little gifts" distributed to the diners. Although none of these matters had yet emerged from the shadows, "at every suggestive phrase or at every allusion which warranted a double interpretation the audience of men in the hearing room roared." Entertainment caterer James Phipps struggled for containment: he denied asking Annabelle Moore to appear only in "stockings and a silk shirtwaist," he denied suggesting that Minnie Renwood should appear as a nude Trilby with "signs or pictures" painted on her body, he denied that he had engaged Little Egypt to appear nude and shackled as "Little Egyptian Slave," he denied attempting to shove a naked Lottie Mortimer out of the dressing room and back onto the stage for an "encore," and he even denied that he was actually an agent. His face scarlet, he denied a great deal more which the newspapers did not print but which made the police commissioners blush. It became clear that Phipps, who described himself as a provider of high-quality entertainment for church events, was actually in the business of procuring, for audiences of rich men, likely girl singers and dancers whom he could then cajole into displaying their bodies and mingling with the audience. Phipps objected to the term *procure*. It also became clear that Phipps was so good at what he did that over the past two years he had stolen from Moore such lucrative after-hours men's-entertainment clients as the Knickerbocker Club.[23]

When Herbert Barnum Seeley, age twenty-five, party host and maker of arrangements, was called to testify, he entered the hearing room with a show of bravado, and "took his seat in the witness chair with a contented little swagger." Seeley was unaware that his friends and dinner guests dreaded his testimony and that lawyers for the defense were certain that he could not, under questioning, maintain his innocuous version of the dinner. Although it was already on record that Herbert Seeley had trailed from one Manhattan agent to the next looking for "real hot" entertainment, under questioning he took responsibility only for Minnie Renwood's Trilby costume. He said he had "expressly stipulated that there should be nothing vulgar in any performance," and when George Liman, an agent at 104 East Fourteenth Street, said that "he had artists who would do almost anything, and asked me if I wanted them with costume or without, I replied that I would have nothing indecent—that I wouldn't tolerate it. I told him that I did not want an exposure of the person." Seeley's purity claims conflicted with his own affidavit; there he had told Police Commissioner Parker that while he had no objection to nude entertainers, it was his settled connoisseur's opinion that full nudity had "too momentary an effect, while if the girls wore gauze drapery the effect would be infinitely greater." Furthermore, as he repeatedly pointed out, the occasion itself required the host to arrange not "a Sunday School" but "something fit for bachelors." In effect, the serenely vacant Seeley tried for containment by suggesting that both he and his activities should be accepted at his valuation of them.

"Young Mr. Seeley," concluded the *St. Louis Post-Dispatch* reporter, "has never had to exercise his brains and consequently his brains are none too active." Seeley's memory continued to falter: he could recall preparing signs to be pinned on Minnie Renwood's Trilby costume but not where he had pinned them, and he had difficulty recalling the nature of the little gifts he had selected for each of the diners. When very specifically prompted, he managed to recall that yes, one of the gifts was a small syringe, an instrument rather like the type he used when he had an earache. The audience, fully aware that the syringe was the most common drug-delivery system of the time, "burst into an uncontrollable fit of laughter." At that point, a party atmosphere fully overtook the hearing room: the *Post-Dispatch* reporter noted that Commissioner Grant, presiding, "seemed to enjoy the testimony very much, and every lawyer seemed pleased that he was there. The trial chamber was thronged with men who rushed out for sandwiches and rushed in with the half-eaten sandwiches in their

AFTER DINNER, WHY NOT HIRE THE POLICEMEN TO DANCE?
[From Yesterday's Evening Telegram.]

hands, grinning policemen, and open-mouthed messenger boys carrying copy to the newspapers."

Every man and boy present in the hearing room had the chance to
attend a vivid oral version of the Seeley dinner, but no newspaper
printed all of the testimony. A reader of one newspaper might acquire
a moderately vivid version of the dinner while a reader of another
might find the door of Sherry's banquet room slammed in his face. A
newspaper might adjure a reader that no one should even think about
the things the newspaper was refusing to report, and then at once suggest that he imagine what those things might be. As the trial heated up,
the *Sun* offered each day many columns that resembled a transcript of
the hearing but whose content had in fact been silently and heavily
expurgated. Hearst's *Journal* splashed its pages with fancifully drawn
swirls of champagne bottles and women's legs but recorded little of

the testimony, choosing instead to repeat frequently the assertion "This material would not interest *Journal* readers." The *Times* chopped its coverage down to only four column inches per day and buried the stories amid the commodities reports on page 15. The *Tribune*, in reporting the testimony of January 8, deployed the term *unprintable* three times within a mere fifty words, and choked on the specifics: "While the occurrences admitted by Mr. Seeley depended for significance upon the construction placed upon his answers or the implication drawn from them, they were of such a nature as to indicate that the now famous, if not notorious, dinner was the occasion of what might be described as extremely 'high jinks.' " On subsequent days, and in the face of further graphic testimony, the *Tribune* lavished its sympathies on Herbert Seeley's "exceedingly unfortunate friends":

> The trial has undoubtedly been a source of much annoyance to Mr. Seeley and those of his guests at the dinner who have been compelled to attend the trial each day. From 10 o'clock each morning until 4 in the afternoon they have been cooped up in a small ante-chamber, uncomfortable and bare but for a few wooden benches. With nothing to distract from the tedious monotony of these hours but their own companionship, they probably felt that they were being amply punished for a night's indiscretion.

For the benefit of any other readers it might still have, the *Tribune* adopted a bored and superior stance, sighing that "the trial of which the public has been forced to hear so much of late dragged its weary way along still further."

Neither the *Times* nor the *Tribune* cared to offend publicly conservative and "prominent" male readers, readers they usually courted, by laying out to them the gaudy details of their own private lives. When the *Times* and the *Tribune* pulled back from the Seeley-dinner story, they were answering not to genteel-respectability codes— even though they sought shelter under that battered middle-class umbrella—but to men's codes of not telling about men's private world; they were protecting not a specific club but a clublike view of men's private group practices. Such protection, however, was both local and spotty. Whereas young women like those who performed at the Seeley dinner had no access to protection, the male dinner guests began to require even greater protection than the newspapers could extend to them. The men had problems on the home front: it was being "whispered around that as a result of the dinner and the subsequent disclosures, discord has prevailed in the homes of more than one of the participants in that bacchanalian feast." W. S. Hart, Chapman's attorney, received unsigned letters demanding that he pose certain searching questions to certain dinner guests, should they take the stand. Hart presumed, from an "angularity" he detected in the handwriting, that these letters came from women yoked in marriage to one or another of the dinner guests. Potential grounds for divorce might appear in the testimony of men under oath answering questions about their goings-on with hired performers, and apparently there were women who wanted those grounds. Lawyers on both sides silently agreed to protect the men and as a result not one of the "young swells and ancient men about town" was ever called to testify.

On January 9, by which point the *Times* and the *Tribune* had banished the Seeley dinner to back-page obscurity, the *Journal* headlined "Public Barred from the Court" and then in effect barred its own readers by offering them portraits of all the participants but almost no information about the testimony. The *Herald,* meanwhile, soldiered on, located "vaudeville doctors" ready to pronounce on the transparency of gauze, and gave the Seeley dinner two full pages of print and illustrations below a stack of headlines.

TESTIFIED
TO THEIR
OWN SHAME

The Women Who Took Part in the
Famous Seeley Dinner at Sherry's
Calmly Told of Exposures
and Other Indignities.

LITTLE EGYPT IN TRANSPARENT GAUZE.

Incidentals were Slippers, Stockings,
a Bolero Jacket and a Little
Band Around Her
Waist.

VULGAR TOASTS AND GIFTS THAT WERE SIGNIFICANT.

Miss Mortimer Says the Straps of
Her Dress Were Cut and the
Men Mauled Her.

CORA ROUTT IN NEGLIGEE.

According to Miss Renwood She Wore Only a Silk Undervest
When Captain Chapman Entered.

GUESTS FONDLED DANCERS.

Spectators at Chapman's Trial Guffawed
And the Lawyers Indulged
In Risqué Repartée.

Each day during the hearing, throngs of men appeared at police headquarters hoping for a chance to see the show and thereby share in the experience of the Seeley dinner. They packed not only the stifling hearing room but also the hallways of police headquarters, and "couriers ran between those within the room and those without, so that when any particularly spicy bit of testimony was given by a witness it was echoed by a roar of appreciation from the hallways." Throughout the hearing—and especially throughout five hours of testimony on January 9—the all-male audience snorted, snickered, roared, and guffawed in an agony of delight over talk of women's bodies; one man was so overcome that he fell into a fit, and the hearing had to be suspended while he was removed, twitching. The *Herald* prefaced each report of a woman's testimony with a detailed assessment of her clothing, hat, hairdo, and overall attractiveness; the *Sun* also assessed the women's bodies as "shapely" and "supple." Although the women were young and showed "some indications of refinement," there was never a suggestion that they were American Girls. In the witness chair, they were "absolutely imperturbable," thus confounding male reporters who expected them to weep like William Moore or blush like James Phipps. Although the *Herald* clung to a notion that any woman speaking in public about men's proclivities must be undergoing "an ordeal," the ordeal was in fact being undergone by others while the women witnesses maintained an icy composure and answered all questions straightforwardly. The women brushed aside Colonel James's prosecutorial attempts to "shield" them and, unlike the men in the room, they did not smile or laugh. The women, after all, knew exactly what a woman's body looked like, knew the names for various parts of that body, and saw nothing funny in any of it; they also knew exactly what men in groups wanted to see and do, and they found nothing funny in that either. They were perfectly prepared to speak of matters that men were quite unprepared to have spoken about. As a result, the male effort at containment had to be taken on by the lawyers questioning the performers.[24]

As the women testified one by one, it emerged that agent James Phipps had specifically asked Minnie Renwood to "expose her bosom," and had attempted to cajole other performers into dancing in the nude because, he told them, it would "make no difference. The men will be all so drunk when you come on that they won't know whether you have clothes on or not." During the entertainment, Little Egypt's "person had been entirely exposed," and the men had touched and pinched her. If a woman had not exposed her breasts while dancing, the men followed her into the dressing room or cut the straps of her dress in order to see her breasts. Lottie Mortimer was the only writer among the women, and although a toast she had written was "thought so good by the gentlemen that they all drank to it and some of them wrote it down in order not to forget it," none of her work was considered printable by any of the newspapers."[25]

Minnie Renwood, who "had made a reputation as Trilby," rose in court to demonstrate that her Trilby costume had consisted of "parti-colored tights" and four signs attached to her body: "Milk below, Furnished by Trilby" over her right breast, "Heart of Maryland" over her left breast, "Held by the Enemy" over her pelvis, and "Secret Service" on her buttocks. Among the Christmas gifts distributed was not only the dubious syringe but also "a little silver thing" whose nature was never specified in print, and Minnie Renwood testified that she had read, as per Seeley and Rich's instructions, a verse to accompany each of the gifts. The newspapers unanimously judged the verses to be "filthy"; only the *Telegram* printed one of them, and at that only the first two lines of the quatrain:

> If she's fretful and ill at ease
> Here's something bound to please . . .

On Sunday, January 11, the *Herald* called in the clergy to furnish some relief by classifying the Seeley dinner under sin and shame rather than under male privilege, social tension, or class division. The clergy obligingly invoked the usual tired comparisons of New York City with Rome, Pompeii, Babylon, and Sodom, denounced "sensuality" in general, and called for "social purity." The *Journal* focused on the Rev. Dwight L. Moody, who not only called for an end to such entertainments but also summoned New Yorkers to a bonfire. Moody blamed art.

> Get all the pictures that are lewd in one huge pile, tear rich paintings from your walls and from the art galleries, collect the newspapers in which are portraits of the nude and make of all of them a huge bonfire in Fifth Avenue. They tell us that we mustn't say anything about art. Bah! I despise these things that are called art. I want to burn these vile things, even if it burns New York. They say I don't know anything about art. No, I don't, not if that is art, and I don't want to.

Nonetheless, the question of what rich men were up to could not be avoided. The Rev. Dr. Louis Albert Banks, pastor of the Hanson Place

Methodist Church in Brooklyn, wanted to know if "the Seeley Dinner is the result of our boasted social advancement and refined culture." He continued, "Think of our great publishing houses and commercial interests in the hands of such men. These men seem to be so far depraved in their tastes that they have not the grace to be ashamed of it." The *Herald* also engaged in furious speculation about what might happen when the hearing reopened on Tuesday and Little Egypt herself took the stand. Although Little Egypt's affidavit had never been officially made public, newspaper reporters knew that she had named certain guests and described their actions in detail. Though they judged her complete statement to be unprintable, they still hoped it might be read at the hearing. Furthermore, they expected great things from her lawyer, William F. Howe of Howe & Hummel, for "when that veteran really lays himself out to do a big thing in the way of furnishing a sensational episode in any trial he succeeds as a past master of the art."[26]

Howe himself indicated that the mere thought of Little Egypt's story as he knew it forced him "to hide his blushes behind his elaborately embroidered handkerchief." For the benefit of his client, however, Howe classified her performance at the Seeley dinner not as sin but as *art*, and then immediately cautioned that it "might or might not be accepted as *artistic*" by others.[27] The *Herald* reporter toted the notion of *art* across town to Little Egypt's agent, and asked if the *artistic* poses planned for the Seeley dinner had been copied from any well-known paintings. The agent replied, "Not much they weren't." In fact, *art*, in the context of the dinner, was unrelated to painting: when Phipps told Little Egypt that the guests at the upcoming Seeley dinner would be *artists*, he meant, in the performance shorthand of the time, that she should appear in the nude. In the beautiful illogic of this code, artists were men accustomed to seeing nude women; therefore men who wanted to see nude women must also be artists. To appear nude before a roomful of artists was to do *art*. As a cover on this occasion, *art* would not suffice, but that was no matter: in the blue-eyed view of white male reporters, lawyers, commissioners, and men-about-town, the Algerian Asheya Waba performing as Little Egypt could be just as readily tucked under the cover of *the Oriental,* with the effect that her performance at the Seeley dinner was judged within a framework entirely different from that surrounding such Westerners as Annabelle Moore, Minnie Renwood, and Lottie Mortimer.[28]

On what turned out to be the final day of the hearing, a crowd of hundreds gathered outside police headquarters and greeted the "chappies" with jeers when they arrived. Inside the hearing room, Little Egypt, wearing a fur cloak and "an intoxicated hat," was just launching into a particularly interesting narrative about what the men did as she danced from table to table and "did a pose to each nice gentleman" when Colonel James, the prosecutor, cut her off. It took, reported the *Telegram,* "all the energy he had to prevent her startling revelations from getting on the record." Hearst's *Journal* shouted, THIS WOULD NOT INTEREST THE JOURNAL'S READERS, and printed a few rows of Egyptian hieroglyphics instead. AFRAID TO GO ON, headlined the *Sun.* The guffawing stopped, the jocular atmosphere dissipated, and the hearing room became utterly silent. The police commissioners bent "grave faces" to a "whispered conversation," and William Howe "leaned over and talked earnestly to Colonel James." There was a pause of many minutes, with Little Egypt still on the stand. While the lawyers and commissioners conferred, Little Egypt removed her fur cloak and "exhibited such an astounding costume," according to the *Telegram,* "that the embarrassment of all was further augmented." The bodice of Little Egypt's dress alternated horizontal panels of black cloth with panels of white lace, and it "fit so close that every curve of the woman's fine figure was perceptible." Men who had earlier hoped to see her in gauze now did not want to see her at all. Deep and silent embarrassment pervaded the hearing room until the whispering men in charge of the show apparently agreed that containment could be achieved only if they, in their turn, donned the guardian-of-public-morals costume. Colonel James, the prosecutor, rose and said to Little Egypt, "That's all. You're excused." The lid slammed shut on the Chapman/Seeley hearing, and the hearing room emptied in a moment.

Although only the grand jury ever heard a full description of the "high jinks" at the Seeley dinner, the cumulative hearing testimony of

January 12 and 13 brought three matters out of the shadows and into view. First, Seeley, Rich, and Phipps had been specifically in search of girl performers who would bare their genitals—called during the hearing "loins," "the area between the feet and the waist," and even "the stomach" in the limited body jargon of the time—and Herbert Seeley had in fact rejected a potential "sensational dancer" when he learned that she did "high kicking, the splits and all the fancy things but wore plenty of lace and didn't expose herself." Second, the Seeley dinner had been a chaotic sexual scene: the drunken guests had grabbed and touched indiscriminately, they had attempted to dance with the girls, they had tried to cut off their clothing, and they had pursued them into the dressing room to watch them remove any shreds of clothing they still had on. Third, and most terrifying to the all-male audience in the hearing room, the testimony had begun to "intimate that the women were not the only persons in the rooms at Sherry's who were careless about their attire." Not only had the spotlight moved from Captain Chapman onto the Seeley dinner but now it was about to move again: a hearing focused on women's bodies was about to refocus on a display of men's bodies. No newspaper was ready to describe the Seeley dinner guests as "naked men," and only the *Herald* edged close to it with the phrase "careless about their attire." The *Sun, Telegram, Times, Tribune,* and *Journal* chose silence.

If clothed upper-class men staring at the breasts of women vaudeville performers was unsurprising, their staring at the pelvis, genitals, and thighs was another matter, and an inexpressible one. If clothed upper-class men fondling those young women was unacceptable, upper-class men shedding their own clothing while looking at and touching naked young hired performers in a banquet room at Sherry's was—for the men in the hearing room—unspeakable. There was no public vocabulary available for speaking of unbuttoned men; men's behavior had outrun men's language for it. That there was in the hearing room a woman, Asheya Waba, who had the language and was ready to use it struck fear into every man present. The police commissioners and the lawyers on both sides of the case agreed among themselves to silence her.

The twentieth century has liked to depict the nineteenth as generally *shocked* by sexual matters. A peculiarity of the *shocked* discourse is that the one who uses the term is never shocked; he or she assumes a superior position, a more knowing position than that of the shocked ones—it is always the stupid and unenlightened other who is said to be shocked. No one in 1897 deployed the false discourse of *shocked*. Male participants in the Seeley affair—those who attended the dinner, those who watched the hearing-room replay, and those who produced the print version—were not shocked but rendered inarticulate in the face of desires for things that could be bought but could not be named. Even the previously intrepid *Herald* scrambled for the abstract high ground, announcing, "What had been mere suggestions in the previous chapters of the trial of Police Captain Chapman yesterday became filthy facts. The deepest depth of the cesspool which has been stirred by the investigation of the bacchanalian revel at Sherry's became actively noisome." Everyone, meanwhile, knew or guessed that there was much more to be said: W. S. Hart, Chapman's lawyer, indicated that "evidence had failed to bring out many things which were done or which had been planned for the dinner." Little Egypt and Minnie Renwood were sent off to complete their descriptions of those "many things" before a grand jury.

The extreme social stress of the Seeley-dinner trial resulted in a frenzy of silence. The few "chappies" who spoke at all expressed mild interest in whether Louis Sherry's business would be damaged but claimed to be otherwise too loftily bored to remark further on the dinner. On January 14, the police commissioners met and laughed heartily over the discomfiture of the "chappies" while maintaining silence about everything else. On that same day, William S. Moore, fifty-two, the man who had instigated Captain Chapman's raid on the Seeley dinner and who had tried to parade himself as a moral crusader, was sitting in his theatrical agency at 10 Union Square when he was "stricken with a congestive chill"; he died four days later. Threatening letters from Seeley dinner guests, it was reported, had frightened Moore into silence, robbed him of his appetite, and left him with a weakened constitution. On January 27, Herbert Barnum Seeley, Theodore Rich, and James Phipps were indicted under sections 168 and 385 of the Penal Code. They were charged with conspiring to induce Little Egypt and Minnie Renwood to commit a crime; the rest of their behavior—all that could not be spoken of—was masked under a second charge of committing a public nuisance.

The rule of silence spread. Oscar Hammerstein, Little Egypt's manager, had been in deep financial trouble until he found he could fill his Olympia Music Hall for *Silly's Dinner,* a "sensational burlesque" featuring Seeley-dinner veterans Cora Routt, Little Egypt, and Minnie Renwood. On January 23, however, District Attorney Olcott suddenly indicted Hammerstein for "maintaining a public nuisance." Little Egypt—in her other lives known as Asheya Waba and as Mrs. Harper—subsequently vanished, but a hundred other dancers calling themselves Little Egypt took her place. On February 4, 1897, the original charges against Captain Chapman were dismissed; a year later, on February 21, 1898, the grand jury dismissed the indictment against Hammerstein. At the same time the grand jury chose to guard the public morals—and simultaneously to reinforce silence—when it dismissed the indictments against Seeley, Rich, and Phipps on the grounds "that to put the case on trial would be against the interests of public morality."[29]

During the months following the Seeley dinner, the police tried to locate and shut down men's private shows; men, meanwhile, avoided any further use of venues as public as Sherry's. They elaborated their secret arrangements for meeting and conducted their amusements in ever more private locations. Not one person who was physically present at the Seeley-dinner ever spoke another word of it in public, but others spoke. In a time when no woman could serve on any jury, Hearst's *Journal* assembled its own "jury of twelve prominent New York women" and asked them for a verdict on the Seeley affair. They agreed in substance that "all men are bad." Fighting back, three of Herbert Barnum Seeley's dinner guests sent out their wives—former American Girls—to make defensive public statements. This rather sorry trio showered both the police and "the dancing women" with scorn. Then, taking advantage of a situation in which no one could publicly name the subject under discussion, they asserted in unison that the entire Seeley-dinner affair had been "a fuss over nothing." It was an odd echo of the remark Little Egypt had made twelve days earlier in regard to the joys of nude dancing: "I do not know why everybody makes fuss." Finally, after begging that their "nervous and excitable" husbands be left in peace, the three women stepped off the stage.

NOTES

1. Howe & Hummel, 19–23.

2. *New York Herald,* 23 July 1899. "Whose Little Girl Are You?" was sung by Catherine Lewis and Henry E. Dixey in Augustin Daly's 1894 production of *7-20-8*. Words by Thomas Naismyth, music by George Rosey. A similar question about young girls pervades the big number performed in *Florodora* (1900) by the Florodora Sextette—the act that brought Evelyn Nesbit to major public notice. Here six "bachelors" seem to admire six Florodoras, but their song asks the question "Tell me, pretty maiden, are there any more at home like you," thus suggesting that they are on the lookout for female items even younger than the Florodoras themselves. *Florodora* was a flimsy but extraordinarily successful English import; after running a year in London, it opened in New York on 11 November 1900, ran for 505 performances, and traveled extensively throughout the United States.

3. St. Denis, 24–25.

4. Throughout most of the twentieth century, Stanford White's behavior remained difficult for men to acknowledge. Here follows a modest selection of the paths male writers have taken. In 1908, Charles Whibley blamed the "blackening" of Stanford White's character on the Yellow Press, cowards, and sentimental spinsters (125–126). In 1922, White's friend Edward Simmons blamed class hatred: "The society of these men was so good, so envied, and so far above the average 'low-brow' that it was preyed upon (probably more than any other class at any other time in the society history of America) by the jealous and the criminal, and blackmailed to an extraordinary degree. [White's group] just naturally broke itself up; Beauty is sensitive and would rather disappear than run the risk of unwelcome publicity" (248). In 1932, Frederick Collins claimed that White was not chasing girls but doing a combination of diplomatic duty and social service: "Stanford White constituted himself a bridge over the gap between the social world of Fifth Avenue and the social world of Broadway. He managed during his lifetime to achieve something very like a community of interest between the drawing room and

the green room. He was well fitted by birth and early training to perform this highly diplomatic task." Collins further shielded White under the always-available covers of genius and businessman: "The impression has been created that Stanford White spent every night of his life for thirty years in pursuing women of the stage and that his relations with them were entirely unconventional. That is absolutely false. He was a genius. He was no persecutor of virtuous young womanhood. He was too busy. He once worked 18 hours a day steadily for a week" (46–48, 181–185). In 1963, Richard O'Connor blamed American aesthetic ignorance for what happened to Stanford White: "Perhaps the most significant facet of Stanford White's personality was the fact that he literally worshiped beauty, whether it was in the hard clear lines of architecture or the softer graces of the human female just a step or two short of womanhood. He was an apostle of beauty in a country which was maddeningly slow to awaken to it. Guilty or innocent of being carried away by this passion for beauty, it was that which destroyed him; his fate encouraged the belief, in paraphrase, that each man is killed by the thing he loves best" (180). In 1983, Leland Roth enlarged White's passion from mere beauty to "a passion for all that life offered, especially the theater," and dismissed his prey as nothing but "pretty butterflies" (331). In 1989, Paul Baker's *Stanny* tore off the veils and brushed aside the excuses in a chapter titled "Bachelors for the Evening" (273–290). In that same year, however, Louis Auchincloss was still asserting that "White's conversations, his parties, his trips abroad seem to blend with his work in a life that was a continuous and joyful pageant. Of course, there were excesses. How could there not have been with such a man? He was not one to be confined to the compound of contemporary morals." Auchincloss blamed the lunatic Harry Thaw for making unseemly revelations about White, represented Evelyn Nesbit as White's sole peccadillo, and denied that White drugged her: "The epicurean in White alone, without assistance from the indubitable man of honor, would have rejected a drugged partner for the act of love" (182, 331). In 1992, paying no mind to Baker's *Stanny,* David Garrard Lowe struggled to shift focus onto White's "religion of friendship," his "penchant for male coteries," and "the *frisson* that came from having a secret place of rendezvous." Lowe further asserted, "There is no evidence that the architect was neglectful of his wife and son. Indeed, he delighted in traveling with them," and commented that Bessie White must have been a woman who "either from nature or from intention, could turn a blind eye to White's male bonding as well as to his other peccadillos" (34–35, 94, 149, 200). One of the rare women to take up the subject of Stanford White's proclivities was his great-granddaughter,

Suzannah Lessard. Perhaps her frank discussion of White and his "sex club" (87ff.) will end the cover-up.

5. Strange, 65–67; *New York Herald,* 4 November 1894, 12 November 1894.

6. Gilfoyle, 285.

7. In *Ladies of Labor, Girls of Adventure,* Nan Enstad takes a revolutionary nonjudgmental view of the spending, appearance, and class aspirations of working girls and analyzes middle-class effort to refuse working girls those pleasures and aspirations: "The notion of character served to distinguish middle-class 'tasteful' consumption from other consumption, and to camouflage the fact that the middle class defined itself in large part through product purchase and use. By taking [the ideology of character] as an accurate description of social relations, historians have missed how the middle class designated the use of certain commodities as superficial, as a fantasy escape, or as immoral, while denying its own commodity consumption" (20).

8. Furniss, II, 101–114.

9. Gordon, 67–72; Bourget, 60–61, 98; Muirhead, 45–48, 58–59; Vaile, 109; Kendall, 205–206; Humphreys, 89–91.

10. Sherwood, 364; *New York Herald,* 21 May 1893,

11. *New York Herald,* 12 March 1893, 5 November 1893, 5 February 1894, 13 January 1895. Candace Wheeler of Associated Artists wrote sympathetically of the "many unhappy and apparently helpless women, dependent upon kin who had their own especial responsibilities and burdens. There was no outlet for the ability of educated women, and yet there was often a pathetic necessity for remunerative work. Women of all classes had always been dependent upon the wage-earning capacity of men, and although the strict observance of the custom had become inconvenient and did not fit the times, the sentiment of it remained. But the time was ripe for a change. It was still unwritten law that women should not be wage-earners or salary beneficiaries, but necessity was stronger than the law. In those early days I found myself constantly devising ways to help in individual dilemmas, the disposing of small pictures, embroidery, and handwork of various sorts for the benefit of friends or friends of friends who were cramped by untoward circumstances" (210).

12. *New York Herald,* 13 January 1895, 1 December 1895, 8 December 1895.

13. *New York Herald,* 21 October 1906. Martha Banta, in *Imaging American Women: Idea and Ideals in Cultural History,* discusses the national fascination with the pure American girl who was "singled out as the visual and literary form to represent the values of the nation and codify the fears and desires of its citizens" and notes that "as the nation increasingly came to see

itelf as a modern world power, its female self-images grew to Amazonian stature" (20).

14. Furniss, II, 101–108; *New York Herald,* 17 April 1893, 26 August 1894.

15. The difficulties of communicating publicly while remaining publicly inarticulate are nicely exemplified by the following *New York Herald* personal ad, dated 9 December 1896: "Sugar and Honey. You add insult to injury. It was I that kept you from being compromised from the beginning, and I will do it to the end of my life you know this. In the other part I was forced, because you would not answer. I am none of those lighthearted creatures that pass from one flower to another. I have protected woman's honor all my life, at the risk of my life, honor and fortune, and, so help me God, I will do it with you as long as there is life in me. I am cured, but feel honor bound to advise you, and therefore ask the privilege of writing one more letter of explanation. Answer to my office that you will give me this privilege, which you owe to me as I must defend myself, and also, if you will, that you forgive my indiscretion when I saw you last. After this you shall not be disturbed any more by me, but I will forever remain your true and devoted friend."

16. "You're Not the Only Pebble on the Beach," words by Harry Braisted, music by Stanley Carter. Published by Jos. W. Stern & Co., 45 East Twentieth Street, New York, 1896. All material on the Seeley dinner is drawn from the *New York Herald,* the *New York Times,* the *New York Tribune,* the *New York Sun,* the *New York Journal,* the *New York Telegram,* and the *St. Louis Post-Dispatch,* 22 December 1896–6 February 1897.

17. When such club entertainments accidentally burst into public view, clubmen either took shelter behind expressions of ignorance and remorse or elected to drop out of sight until the storm blew over. In March 1893, the Hanover Club, the "swell" social organization of the Eastern District, Brooklyn, held a "stag racket" with "novel features in the shape of slugging by negro pugilists and high kicking by fleecy-skirted lights of the vaudeville stage. None but the members of the Entertainment Committee knew what was coming," it was asserted, when the Bellfield Brothers appeared on the stage. "They are colored youths and were stripped to the waist when they tripped out upon the stage with three-ounce boxing gloves on their hands. Clever leads, counters, upper cuts and swings elicited no applause from the horrified Hanoverians, and when finally one of the Bellfields punched his brother off the stage there was an uproar of condemnation which put an immediate end to the contest." Then the Hengler sisters took the stage in short skirts of the "lightest and fluffiest texture. During their dance their toes most frequently pointed roofward. Some of the members that saw them have admitted that they lay

awake when they got home, thinking of what the women folks would say when they heard of it. Now the members of the committee that handled the Tuesday night 'stag racket' are nowhere to be found" (*New York Herald,* 2 March 1893).

18. *New York Journal,* 3 January 1897. In *Steppin' Out,* Lewis Erenberg discusses women's enlarged access to public entertainment in the 1890s, and in *Going Out,* David Nasaw documents the stunning increase of public amusements during the same period that groups of men began to arrange private shows. Paul Baker's *Stanny* gives a good view of the linguistic codes upperclass men used to signal each other that a special show would occur. In the context of the Seeley dinner, *Complete Bachelor Manners for Men* (1896), interestingly enough, notes that the bachelor dinner is "the most important social function of the wedding week," but says nothing about entertainment and is firm in its belief that "a dinner with favors is not permissible in these days" (181–183).

19. In 1895, the sixteen-year-old Annabelle Whitford did her serpentine dance and her butterfly dance before Thomas Edison's kinetoscope at his Black Maria movie studio in West Orange, New Jersey; many New Yorkers undoubtedly watched that performance at Edison's public viewing room in Manhattan. The few surviving seconds of Annabelle dancing can be seen on videotape: *The Art of Cinema Begins,* Video Yesteryear, Sandy Hook, Connecticut, 1999. Although Annabelle swore at the hearing that she had never once appeared without tights, on the Edison film Annabelle is obviously barefooted and barelegged, thus playing to one of the greater erotic interests of the 1890s.

20. Newspaper coverage of the Seeley dinner had immediate effects on other such bachelor-life events. On 23 December 1896, the *Herald* reported that women in Mount Vernon, New York, were surprised to learn that "one of the Sherry dancers, 'the Kentucky thoroughbred' Miss Cora Routt, cavorted before the New Rochelle Rowing Club's smoker. When the wives of the members of the rowing club read of the Sherry raid, they jumped to the conclusion that the club's smoker was a similar affair, and several prominent men and their wives are said to be 'on the outs.' When I asked some of the members of the club about the smoker, they declared it was genteel and respectable and good enough for a Sunday school performance. They say it glibly as if they had been repeating it to their wives nine hours out of every ten for the last three days. One man appeared before a notary public this morning and made an affidavit which contained the declaration that the smoker was perfectly respectable in every respect. This was for presentation to his wife."

21. In less than a month, Mrs. Bradley Martin would make the identical

mistake and give out the names of her guests and their costumes to the newspapers.

22. In *Howe & Hummel: Their True and Scandalous History,* Richard Rovere records Howe on the subject of his passion for diamonds: "In New York, no one would have considered hiring a trial lawyer who didn't sport a few diamonds. I really didn't care much for them then, but now I love them. I love the light they throw. I can't get enough of them." Howe, according to Rovere, "wore diamonds on his fingers, on his watch chains, as shirt studs, and as cuff buttons. He wore neckties only on occasions such as funerals and hangings. In place of a tie, he generally wore one of a number of diamond clusters from his large collection" (17).

23. In "Clubs and Club Life in New York," Robert Stewart identified the Knickerbocker as "undoubtedly our most exclusive fashionable club. You may always be gratified by seeing some of the very greatest swells seated in its square bow window at Thirty-second Street, and it is as near an approach to the Travellers and the Marlborough, of London, as it is possible to attain in this deplorably democratic land. Mere membership is a passport to society, and there are men who would give a hundred thousand dollars to be able to engrave its aristocratic name in the left hand corner of their visiting cards" (109).

24. In the 1890s, men grew increasingly interested in women's clothes and in effects that might be achieved by wearing them. Even as women "renounced and anathematized the corset," men began to wear corsets. The "combined belt corset," its promoters claimed, "showed off the figure and enabled the tailor to ensure an effective fit and distinguished appearance. It is a necessity to most men for the promotion of health and comfort, together with an upright, soldierly bearing. It expands the chest. It supports the spine and holds the figure erect. It protects the lungs and kidneys from cold. It supports the stomach." The *Herald*'s own haberdashery expert found such remarks "equivalent to saying that the very thing which its opponents contend it doesn't do for women it will do for men. That may be labelled 'important if true.' Later on they may try skirts. This is an age when things are getting turned around generally, and he would be a rash man who would venture to predict where it will all end" (*New York Herald,* 11 November 1894).

25. For a woman to speak or write of sexual matters could be downright dangerous. In 1895, a woman under twenty who "figured prominently in society" in Rhinebeck, New York, offered to one Dodd, a nineteen-year-old male admirer of hers who lived at 8 Broome Street, a poem of "alleged shocking tendencies" and some prose pieces which were also "of erotic tendency." Dodd showed the manuscripts to a friend in New York City, who passed them

on to a printer, Louis Force. Force in his turn "wrote to a fifteen-year-old boy in Rhinebeck offering him the agency for a proposed publication which was to include the writings of the Rhinebeck belle. The mother of this boy found the letter and sent it to the Society for the Suppression of Vice. An agent of the society caused Force to be taken before United States Commissioner Benedict, in Brooklyn, and he was held to answer a charge of having sent improper matter through the mails." Men passed the Rhinebeck writer's work from hand to hand and judged that "while correct in grammar and rhetoric, her manuscripts contained alleged depraved sentiments." Then they sought either to make her writing more public or they opted for the censor's classic move: they read the stuff themselves and then moved at once to conceal it from other readers and to condemn the writer. The woman writer herself was said to have fled the state in alarm (*New York Herald,* 8 December 1895).

26. In *Ziegfeld Girl*, Linda Mizejewski discusses the fantasy division between the American Girl and Other Girls, a fantasy that Florenz Ziegfeld intensified in numerous magazine articles about how he "picked" girls for his spectacles. Ziegfeld regularly "explained that his Follies Girls were 'native' Americans whose grandparents and great-grandparents had been born in this country. The point of this outrageous fabrication was that Glorified American Girls were not supposed to be recent immigrants from southern and eastern Europe, hence not ethnic, dark-skinned, or Jewish" (8).

27. The *danse du ventre* traveled from Chicago's Columbian Exposition to New York City, in which venue it was sometimes legal and sometimes illegal. At the time of the Seeley dinner, it was legal; Little Egypt was performing it onstage every night of the week. Howe & Hummel liked to defend *danse du ventre* cases, and were expert at turning such events into chaos. In regard to one such case in December 1893, antivice lord Anthony Comstock announced that "any man having respect for womanhood would condemn such a performance." Comstock was "of the impression that in their own country the dancers would be beheaded." Abe Hummel, the dancers' lawyer, asked Captain Berghold, who had arrested the dancers, if he knew which part of the anatomy was the stomach, a crucial location for performance of the *danse du ventre*. Berghold said no, he did not know, and he was not a doctor. Then, because the judge would not allow the dancers to put on a demonstration for his assessment, Hummel trapped the arresting officers into demonstrating the *danse* themselves. Captain Berghold, before leaving the witness chair, was made to "give a feeble imitation of the contortions which he had considered ground for the arrest. The Captain lacked abandon, but Dennis McMahon, the ward man who succeeded him on the stand, after a brief statement about the unseemly 'extorting' of fair forms, which elicited an exclama-

tion of horror from Lawyer Hummel, illustrated the dance with greater freedom. He stood up, pulled the skirts of his coat out of the way and executed a slight twist. Fatima laughed outright. Later Captain Berghold climbed back into the witness chair unsummoned. He wished to explain that he had rambled through anatomy and had finally trailed the stomach to its lair. He wanted to state definitely what he had meant by 'stomach' in his previous testimony" (*New York Herald,* 4 December 1893).

28. The Larchmont 22 were a dubious group in other ways. On 14 January 1897, the *Herald* reported that "the trial of Captain Chapman has created a sensation in Larchmont. Several of those implicated in the Seeley dinner belong to the righteous faction of the Larchmont Yacht Club, which recently caused proceedings to be brought against Mrs. May Charman, owner of the Victoria Hotel, alleging that she sold liquor during prohibited hours. The developments of the Chapman trial have been balm to the feelings of the unregenerate element in the Larchmont club, who have found the restraint imposed upon them irksome."

29. *New York Times,* 22 February 1898. In *Looking for Little Egypt,* Donna Carlton offers numerous photos of numerous Little Egypts, one of whom may have been Asheya Waba.

MAY BE AN ANARCHIST.

CHAPTER 6

Incurable Panic

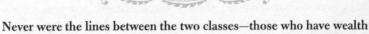

Never were the lines between the two classes—those who have wealth and those who do not—more distinctly drawn. Whether we like it or not, it is an incontrovertible fact that a large portion of our population is discontented, and does not hesitate to express its feelings.

—William Stephen Rainsford, 23 January 1897

Why should we tolerate conditions that breed millionaires and paupers?

—*New York Journal*, 29 January 1897

AT HALF PAST FIVE on Thursday, November 16, 1893, Delmonico's at Fifth Avenue and Twenty-sixth Street was unusually crowded for that time of day; fashionable men and women who had just left the annual Horse Show at Madison Square Garden filled every available table. Outside the restaurant stood George Roeth of 530 West Forty-sixth Street, a well-built, sharp-featured, twenty-eight-year-old stonecutter. On November 10, Roeth had quit his job with the Dock Department; two days later his mother gave him some money and told him to find a furnished room because he was "too quarrelsome to live at home." On the 16th, after wandering Manhattan for most of the day, Roeth had come to a halt at Delmonico's and was leaning across the iron railing that surrounded the restaurant. Looking through Delmonico's Fifth Avenue windows, he trained his eyes—"glaring light blue eyes" if you read the *Herald* or "glittering black eyes" if you read the *Times*—onto the "brilliantly lighted" scene within. The plate-glass windows delivered their usual double social message: some swells were inside but George Roeth was outside.

After standing quite still and watching for some minutes, Roeth suddenly drew a revolver, brandished it above his head, and shouted, "Curse the rich! Curse them now and for all time!" Then he leveled the revolver at the restaurant and fired through one window after another. One bullet whizzed past a waiter's head. Another streaked across a table at chest level. Outside, pedestrians scattered and cab-drivers jumped for cover in their vehicles. Still firing and shouting "Down with the rich!," Roeth dashed through Delmonico's main door. Inside the dining room, "terrorized ladies crouched behind pale-faced escorts" and from beneath a table Manager Garnier "filled the house with trumpet-toned cries of 'Murder.' " To a man, the waiters discarded their usual hauteur and became acrobats, leaping over chairs and patrons to reach the windows opening onto West Twenty-sixth Street. "Unfortunately," reported the *Herald*, "only two of these windows were open, and into them the terrorized crowd securely wedged itself—separate masses of kicking feet and waving arms."

Roeth had fired at the ceiling and was absorbed in watching plaster shower down when the small but "plucky" George Hancock of 100 West Eighty-sixth Street sprang at him. Hancock was no match for the powerful Roeth but was soon aided by Felix Jewell, engineer of Fire Engine No. 16, who had been loafing outside the restaurant, and by park policeman James Dillon, who had been on duty in Madison Square. Both Jewell and Dillon later reported that "at the moment of the capture there was not a single employé of the place in sight beyond the fear-maddened waiters wedged in the windows." Escorted by a crowd of hundreds, Officer Dillon marched Roeth off to the West Thirtieth Street station; along the way, Roeth explained to Dillon and to the crowd in general that "he had only intended to frighten the people whom he saw in the dining room feasting like Kings, while there were thousands of poor workingmen going about hungry without a cent in their pockets. 'Those people in the restaurant,' he said, 'never earned a cent in their lives.' " At the station, Roeth laughed intermittently while explaining himself to Sergeant Lane: "You see, Sergeant, I don't like to see the rich people enjoy all the blessings of life while the poor starve. I did this shooting tonight with the idea of frightening them into a change of heart, don't you see?"

If the poor man wanted to have his say about America's social situation, the rich man wished to utter nary a word. Back at Delmonico's after Roeth's attack, the swell diners fled: "They refused to give their names to the police and used every means to conceal their identity. One fine-looking man, accompanied by two ladies in elaborate dress, bundled his party into a cab at the crowded entrance a few minutes after the shooting, and, as though determined to balk the curiosity of bystanders, cried, 'Drive me to Wall Street!' " If the rich and those who catered to the rich feared death at the hands of a would-be anarchist, they had an almost equal fear of newspaper notoriety. With no way to confront the social tensions of 1893, the rich ran and hid from threat, while those who served them silently pretended that there had been no threat. By half past seven, a version of order had returned to the restaurant: Charles Delmonico had ejected the reporters, Manager Garnier had lost his tongue, and the waiters had regained their hauteur. George Roeth was in jail and Delmonico's dining room was full again.

On November 17, Charles Delmonico sent his bookkeeper, Simon Weir, to Jefferson Market Police Court to tell the story of Roeth's behavior; no other complainant appeared. An "agitated and trembling" Roeth explained to Justice Hogan that he had had a "great deal to drink" and had "a lot of cussedness" he was wanting to rid himself of when he "got along to Delmonico's" and saw it full of rich people enjoying themselves. "I just fired into the place on general Socialistic principles. I had the dining room all to myself for a while, and I certainly had more of Delmonico's than I ever expect to have again. I never intended to kill anybody." Justice Hogan asked Roeth not a single question before committing him to Bellevue Hospital to have his mental condition examined.[1]

Nothing better illuminates the world of economic failure and personal catastrophe in which George Roeth took his shot than a small but representative selection of the hundreds of panic headlines that New Yorkers read in the *Herald* between late April and the end of December 1893. During this eight-month period the story of the greatest depression to have thus far hit the United States gradually traveled from the back-page financial columns to page 5, where it challenged the Bradley Martin wedding story in importance, and onward from there to the front page, where it took up residence. Stories about a new world of failure, fear, fraud, and incompetence, however, continued to vie with stories about a still-ongoing world of wealth and display. Each day's newspaper was a circular high-tension contest among business failures, party favors, defalcations, luxury displays, bank closings, trips to Europe, and starvation.

April 30	Wall Street's growing gloom. Money is tighter everywhere.
May 5	Wild time in Wall Street. Fortunes lost in a day. Fears are entertained of further embarrassments. Speculators caught in a general slump.
May 7	Wall Street's biggest crash.
May 13	Country banks carried down. Local banks still unhurt.
June 6	Almost a panic in Chicago. Frightened depositors crowd into banks and withdraw their accounts.
June 7	Canal Street Bank closes. Depositors knock in vain on the closed doors.
June 14	Irving Savings Bank accounts short by $70,800. Panic in western cities. Runs on all the Omaha banks. Business houses suspend. Financial contagion overtakes Kansas City.
June 29	Silver goes tumbling down. Washington authorities will let things take their own course for the present.
July 3	Glad tidings for the poor. Seven *New York Herald* free ice dispensaries to be opened.
July 9	Financiers at summer resorts show great anxiety. Roast beef and mashed potatoes instead of woodcock and romaine salad. The women—dear creatures!—as gay and beautiful as ever.

July 19	Down go more Denver banks. Sixteen failures in one day. Heavy losses in Kansas. No local failures reported.
July 21	Fifteen thousand men discharged by the railways, mines, and factories making their way to Denver.
July 23	Wall Street's hardest week. A fall so sharp as to inspire hope that the panic has culminated.
July 28	Hard for the workingmen. New England factories and Pennsylvania industries and Michigan mines closing or preparing to do so.
July 30	Big rush for Europe's shores. Nearly a quarter of the tourists who sailed yesterday are New Yorkers.
August 2	Fortunes lost in Chicago. John Cudahy's millions gone. With many others the daring speculator was ruined within a short half hour. Taught a lesson by Armour.
August 5	Thousands of men out of work. Industrial depression all over the country. Fifty thousand in Chicago alone are idle.
August 7	The *Herald*'s first banner headline: RESTORE PUBLIC CONFIDENCE.
August 9	One more New York bank likely to close—the Madison Square.
August 13	Wall Street's waning trade. Things are all topsy-turvy. Speculation defies all ordinary rules. Wages to be reduced.
August 17	Paraded to show their poverty. Spontaneous demonstration of unemployed men who had long sought work. No bands or leaders in line. Many trades and nationalities represented. Want work, not charity. Resolutions denouncing moneyed men adopted at a mass meeting after the parade.
August 18	Rioting by idle workmen. Police reserves called out to disperse desperate crowds in the East Side. Frenzied demands for bread. Blame put upon monopolists.

August 19	Anarchy publicly preached. Action urged by Emma Goldman. If those who are hungry cannot get bread they must take it. Bold promises cheered.
August 20	Charming ball at Mrs. Burden's. Newport society delighted with the decorations and the perfect arrangements. Very handsome German favors.
August 25	Madison Square Bank's big loss. Surplus wiped out. Other officials borrowed. Their names kept secret. Large blocks of worthless stock in the assets.
September 3	Joys of summer life. Connecticut mill operatives find comfort and pleasure at small cost. How families of five persons live on seventy-five cents a day and enjoy themselves.
September 4	Hard times fill the navy.
September 7	No hope left for many bank depositors.
September 9	Examiner makes a secret of bank affairs. Information refused of Madison Square Bank's queer workings. "Not the Public's Business," he says.
September 11	There may come a big explosion.
September 16	Dishonesty sweeps like a wave of crime over all the land.
September 19	Arming to fight want in Yonkers. Six thousand people idle. Many families have no food. Outlook for the winter causes apprehension.
October 4	Stock market virtually at a standstill.
October 14	Four hundred and six failures reported in the United States during the week, against 365 in preceding week.
October 18	Time for disappearing. These merchants, whose affairs are involved, are reported to have dropped out of sight.
October 24	Madison Square Bank officials under arrest. President Blaut, cashier Thompson, and directors. Charge of forgery. False statements made.

October 25	Four more Madison Square directors arrested. Fraudulent insolvency. Enormity of forgery and perjury.
October 26	Was the Madison Square institution bled? Political influences said to be working in favor of the arrested directors. The men may be shielded. Suggestive friendships that existed between the officers and men high in politics.
November 5	Veterans cashing their pension checks. Men who fought for their country enduring an old age of poverty.
November 17	Pistol bullets in Delmonico's. Crying "Down with the Rich!" Crank George Roeth fires through the windows. Confusion reigns supreme. The waiters and richly dressed diners rush pell-mell for safety.
November 19	Extravagances of an age of luxury. Costliness of high living exemplified in present social conditions in New York. Growing wealth and display. Extravagant tastes finding gratification in gorgeous hotels, dazzling restaurants, and expensive club life. Where money is king.
November 24	Indictments for outsiders, too. They played an important part in the downfall of the Madison Square Bank.
December 3	Pay your money, take your choice. All the good things of life to be bought at the retail shops. Merchants welcome the merry yuletide season.
December 4	Michigan miners starving by thousands.
December 6	Income tax a death knell. A landslide of disapproval. Many sound reasons given. Unanswerable arguments why such a tax is both obnoxious and impracticable.
December 7	Mayor Gilroy expresses doubt. Knows of no present means to cope with the evil or prevent widespread suffering.

December 11	Money for Chicago's poor. Unemployed estimated at 117,000 in Chicago. Such a calamity as is now with us has never visited Chicago before.
December 16	Inheritance tax to be adopted by Congress. Not burdensome on anybody, easily collected, and would bring in a large amount.
December 17	Digging for their lives. Long Island families depend upon woodchucks for subsistence. Sources of income cut off and it is hard work to ward off starvation.
December 31	How the bank was smashed. Inside history of the peculiar methods which wrecked the Madison Square Bank. "Gosh! We're busted," said Director Uhlmann on the Saturday before the closing.

On October 28, 1893, just as the World's Columbian Exposition in Chicago was being dismantled, Chicago's Mayor Carter Harrison was shot to death in the front hallway of his own home by a disgruntled job-seeker named Prendergast, a man thereafter always referred to in newspaper headlines as Crank Prendergast. This singular and shocking event allowed standpat social commentators to label as a "crank" almost any American who demanded that his miserable situation be publicly addressed. They appealed to perpetually unsound but nonetheless popular imitation theory when they insisted that it was the assassination, not the general social situation, that was bringing to center stage some very dangerous people. The Harrison assassination also became a useful locus of blame for other effects of the general social dislocation of 1893. On November 1, 1893, the *Herald* tied all fear and social chaos to the Harrison assassination:

CRANKS EAGER TO TAKE LIFE.

Prendergast's Awful Act Has Given Rise to an Epidemic of Insane Murderers. Most of the Army of Lunatics Appear to Be Money Mad—Six Women Are Among Them.

According to the *Herald*, individuals "hitherto regarded as harmless, although eccentric" had interpreted the Harrison assassination as "the signal to engage in murderous enterprises." From there, it was but a small step to the point of transforming into a murderous lunatic any "harmless eccentric" who talked about money. To ask another human being for money signaled that one was lodged in a delusion about how the world worked—and that one was thereby suspect, sinister, or dangerous. In November 1893, when a man stepped directly into the path of New York's Mayor Thomas Francis Gilroy and said, "You're a dude and a plutocrat and I want a dime," the question to be settled was neither the nature of the situation that had put him in need of a dime nor whether the Mayor should give him a dime. The question was whether he was a dangerous crank or merely a drunk. To demand money was so socially inappropriate, so out of tune with the times, that a man who made a demand was suspect, while a man who refused the demand escaped judgment. The problem lay with the individual who asked, not with the individual who refused.[2]

Around noon on April 30, 1894, Camille Rheinhardt, twenty-six, a "tall cadaverous man with long yellow hair and whiskers, walked briskly through West Fifty-fourth Street and halted in front of No. 4, the handsome residence of John D. Rockefeller." Rheinhardt had an "unhealthy pallor" and those "glittering eyes"—gray ones this time—that newspapers always ascribed to the crank. He rang the doorbell, and when Christopher Olsen, Rockefeller's head butler, opened the door, Rheinhardt said: "I am here to see Mr. Rockefeller. The fact has reached me that Rockefeller has to his credit a matter of something between thirty and forty millions. I have called for my share, and I propose to carry it away with me."

When Olsen claimed that Rockefeller was out, Rheinhardt called him a "flunkey" and demanded to see the ladies, who he was sure were in. At that, Olsen slammed the door; Rheinhardt took a seat on the steps. Rheinhardt never displayed a weapon or made a threat, but during the next six hours, he did periodically ring Rockefeller's doorbell and shout, "I'm waiting for Rockefeller. You can bet I'll see him," and "Rockefeller must see me. Forty millions is more than any one man made." To these remarks, Olsen the butler, from his side of the door, responded by shouting, "Get out."

At six in the evening, the besieged Rockefeller women ordered the "quaking" Olsen to go for help. Unlike the Delmonico's waiters who jumped out the windows while George Roeth fired at the ceiling, Olsen did not have the option of flight, so he "reluctantly buckled on the armor of duty and proceeded to show what an amount of self-sacrifice a high priced servant is capable of in a pinch." Olsen stealthily left the house via the basement door and darted off to the East Fifty-first Street station. Within five minutes, the police had Rheinhardt in custody and Olsen the butler had lodged a charge of disorderly conduct against him; no Rockefeller ever became involved. After Sergeant Staincamp searched Rheinhardt and found that his pockets contained "absolutely nothing," he was judged "evidently insane."[3]

Suffering Americans were, it seemed, socially forbidden to discuss their suffering with those who were not suffering. Some Americans wanted from the rich not money but the know-how they needed to make money; at the time, however, a poor man looked to be downright crazy if he tried to acquire financial know-how directly from a rich man. In 1893, neither dimes nor the secrets of acquiring dimes were to be shared. In October of that year, Margolia Andrews, thirty-four, a railroad telegrapher, left behind a family of three in Kansas City and went to New York to look for work. He took a room in a boardinghouse at 378 Pearl in Brooklyn. There, the stock market was Andrews's sole subject of conversation; he "watched the stock quotations closely, and when noticing the advance in some security he would remark to the other boarders: 'If I had only put $5 on that stock I might have made some money.' " Unable to find a job, Andrews eventually abandoned the search and turned instead to haunting the Missouri Pacific Railway offices of Jay Gould's sons, George and Edwin, at 195 Broadway. Day after day, a shabby and dazed-looking Margolia Andrews drifted into the Goulds' outer office and asked to see George Gould in order to obtain from him a few tips on the stock market; day after day Andrews was told that Gould was "out." As weeks went by, Andrews's condition worsened; he was clearly malnourished, possibly even starving, and he wept continually.

On October 30, he demanded that a note be carried in to Edwin
Gould.

> Mr. Edwin Gould: Will Western Union go to 100? I want to work
> the bucket shops? Answer at once. Andrews.

Edwin Gould read the note, called Police Superintendent Thomas
Byrnes, and requested that Andrews be arrested. "The man looked
perfectly harmless," said Gould, "but I believed that it would be better
for him and better for us if he were taken care of." At police headquarters, Andrews was found to have a quarter in one pocket and a bottle
of medicine in the other. He had no weapon but he did have an explanation: "I have been following up quotations in stocks, but I am not so
experienced as he [Gould] is at the business. He could give me a good
tip if he wanted to." Having articulated the one incontrovertible truth
that he had a grip on, Andrews was locked in a cell.

The social treatment accorded to Rheinhardt, Andrews, and hundreds like them was validated when, in the months following the panic
of May 1893, important experts stepped forward and dispensed sage
advice on dealing with—or disposing of—such men. The prominent
Professor Landon Carter Gray, in a paper read before the Society of
Medical Jurisprudence, 17 West Forty-third Street, defined cranks as
"turbulent lunatics" who "come to the surface at every public commotion." Considering what most "cranks" had to say when they were
arrested, however, he might more accurately have defined them as
Americans who publicly expressed social ideologies unacceptable to
the dominant group and as persons who acted out their ideologies in
social settings with social consequences. In all of Professor Gray's
long paper, however, there appeared no mention of economic depression or extreme social stress; the catastrophe of 1893 he veiled under
the phrase "public commotion"; then he asserted that only singular
troubled individuals had been touched by it. Gray further warned that
cranks had learned, from studying anarchist leaflets, how to handle
high explosives; he suggested arresting "every case of chronic insanity" in New York and holding all of them until they could be properly
examined by experts such as himself.[4]

With experts like Landon Carter Gray eyeing the local "crank" population as a potential career opportunity for themselves, an American asking another American for the basics of survival had to calculate the conditions under which he would ask. A successful beggar had to protect not only his personal self but also the person from whom he would beg. Beggars began to work not face-to-face in daylight but rather under cover of darkness, or from across a protective wall; even under those conditions, they learned to ask not for money but for leftovers, discards, or used-up stuff. Starving persons took to crouching in darkness on the Broadway side of hotel cafés and restaurants; there they selected an open window and awaited the opportunity to ask an individual diner for any food that he seemed unlikely to eat. Their timing had to be exquisite: they had to ask just as a diner gave signs of being finished but before a waiter observed the same signs and came by to clear the table. Meanwhile the servitors inside the restaurant had their own act to perform: they were required to protect the well-fed from the unfed by making at least a threatening gesture at the beggars lurking outside. They usually managed to be just a bit behindhand with the threat. On November 3, 1894, a *Herald* reporter had just finished his meal in a hotel restaurant and had lit his after-dinner cigar when he was startled by "a voice out of the darkness."

"Hey, mister, won't you please give me that bit of bread?" said the voice, in a stealthy sort of stage whisper. Then the pinched face of a boy became visible in the light that shone from the brilliantly illuminated restaurant. It was a hungry face as well as a hungry voice. I slipped the bread out to the lad, and he scurried away, just as the head waiter caught sight of the face at the window and started toward my table threateningly. I had scarcely laid down my evening paper and resumed my cigar when there came a second appeal out of the semi-darkness of the street, and glancing out of the window again, I realized that I was once more being accosted. "Say, mister, won't you give me that newspaper?" was what this voice said. "What do you want with it?" I asked. "Want to sell it," the voice replied. I gave up the newspaper as willingly as I had given up the bread and then called for my check. The head waiter followed me half way to the door, making profuse apologies for what he pretended or fancied to be the serious annoyances to which I had been subjected.[5]

After the crash, uncountable numbers of men set out for America's big cities in the belief that there they could surely find some work. A job-seeker could expect more sympathetic social treatment than that extended to men who demanded a dime or begged for bread, but he had no greater access to social aid than did cranks and mendicants. In February of 1894, John Lyons, a blacksmith from Canton, Massachusetts, a father of three who had lost both his shop and his home, walked to New York City to look for work. He had fourteen dollars in his pocket, and the walk took him and his longtime employee Cornelius Buckley four days to complete. For two weeks and without success, Lyons looked for work in New York City. Then his money was gone, and he and Buckley set out to walk back to Canton. Because they were weak from lack of food, their progress was very slow. At Kingsbridge, "a kind-hearted citizen gave them a few cents to buy some bread and directed them to the police station, where they were given lodging for the night." Toward morning, John Lyons began to scream and weep uncontrollably; he cried out that his wife and babies were dying, and that two men were trying to kill him. When he was brought before Justice Feitner in Morrisania Police Court to tell his story, he was incoherent and "he stared vacantly around the room." Justice Feitner pronounced Lyons to be "showing every evidence of insanity," and then addressed his companion, Buckley: "What are you going to do to keep from starving?"

"I don't know," replied Buckley.

Justice Feitner advised Buckley to apply to charitable societies for assistance, but Lyons he committed to Bellevue to have his sanity evaluated. Clearly Lyons understood what was happening to him, and just as clearly he did not want it to happen. As he was being removed from the courthouse, he broke away from his keepers and threw himself violently forward down the stone steps. At the Harlem Hospital it was discovered that Lyons had broken his back, and "the doctors said that he could not recover."[6]

When insanity came to be represented as an existing but concealed state, a widespread condition just waiting for an individual trigger, it looked nearly unavoidable. The newspapers in 1893 and 1894 stuffed column after column with reports of "crank behavior," lunacy and insanity. They effectively painted a daily picture of human insanity, and offered running commentary on the visible signs of insanity, the rules governing sane behavior, and the punishments awaiting those who failed to appear sane in public. These same "crank" stories, however, also operated to define their social obverse: an approved human being that Charles Rosenberg calls "the ideal social type." To meet that ideal, to appear sane, a person had to demonstrate near-perfect public self-restraint. A sane person never accosted strangers or launched public displays of loss or fear. To be sane was to be quiet, undemanding, self-sufficient, and bereft of social ideas; to be insane was to scream out rage, despair, and entitlement, to ask others for help, and to have ideas about the social order.[7]

In popular insanity-lore, a business-panic model took hold of the human psyche and made it mirror the terrible events, fears, and losses of the time. Just like a person who lost a fortune on the stock exchange within half an hour on May 6, 1893, a person could lose his or her sanity suddenly and irreversibly. The loss could be triggered by a shock or even by a surprise; all shocks came to seem like big shocks, and all surprises were bad. Furthermore, just like the apparently hopeless state of the economy in 1893 and after, insanity was adjudged irremediable, an awful and hopeless condition from which there was no recovery. Annie F. Farrell, for example, was in the audience at the H. R. Jacobs Theater on Third Avenue when, in an admired special effect of the time, a fire engine sped across the stage. Taken by surprise, Farrell began to scream and could not stop; she was arrested, tossed into Bellevue, and judged to be "an incurable case." In a similar case, Mamie Lynch, a sixteen-year-old drop-press operator, went off to eat lunch one day, and while she was away her coworker, the dangerously prankish Etta Jamison, concealed a four-foot-long rubber snake behind Mamie's press. When Mamie resumed work, Etta tugged on a string that made the rubber snake seem to be creeping out of Mamie's press.

> Mamie shrieked and turned to run. Etta coiled the snake in
> folds about her own neck and ran after Mamie screaming in

pretended fright. Mamie turned her head and seeing Etta, as she supposed about to be strangled by the monster, broke into a peal of blood-curdling laughter and swooned. She has been laughing and crying by turns ever since. The fright has had some peculiar effects upon the brain which the physicians are unable to determine. It is feared her reason will never be regained.[8]

In a further mirroring of the financial losses of 1893, heavy human involvement in any activity came to be looked on as a potential trigger, rather as if one had invested everything in Northern Pacific Railroad stock or squirreled everything away in the Madison Square Bank, only to have 1893 wipe out every last penny. Florence Allison, "crazed by a passion for fine clothes," was judged insane when she tried to protect her wardrobe from thieves by putting on all of her dresses at the same time. George Simonds, a clerk, "thought of money until he went mad" and tried to give away to the poor two million dollars that he did not have. Too heavy an investment in religion, for example, could make people loud, and loud meant crazy. Helena Fetter, who "attended religious services daily for a week, had her mind affected by them." She began to shout out religious texts in public, and was "locked up in the insane pavilion at Bellevue, suffering from what the physicians think is religious mania." When Charles Montz suddenly bounded into the pulpit of the Church of St. Vincent de Paul and yelled "I want bread!" he was diagnosed as "a madman crazed by bible study." Conversely, loud antireligious talk could also brand the talker as a "dangerous crank."

The insanity concept engulfed not only every shade of eccentricity but also many everyday human activities and small pleasures. Middle-class reformers managed to link their antipleasure ethic with insanity diagnoses: to play cards too frequently or to smoke too many cigarettes, they asserted, caused insanity—as did following a sport too intensely. In November of 1893, the newspapers began to call lower-class baseball fans "ball cranks." "Wall Street men," however, escaped that designation; no matter how vigorously they screamed and pummeled each other in the Brokers' Box at the Polo Grounds when the Giants won, they were "not cranks, but part of the great throng who love baseball."

Football fans were in significantly greater mental danger. Dennis Dean, an Atlantic City hotelkeeper, became "insane over football."

> Dennis Dean has been for a long time wildly enthusiastic over football, and on Saturday his friends observed that he acted queerly. In the afternoon his mind gave way entirely just after a heated discussion over the game. He rushed into the Good Will Hook and Ladder Company's house and seized an axe from the truck, which he flourished threateningly and tried to strike the employes. He was disarmed after a struggle and allowed to depart. Dean then secured a pistol, and going into the saloon of Thomas Kilcourse, attempted to blow out his brains. He was then taken into custody by a policeman and put under lock and key. City Physician Ulmer examined him and gave it as his opinion that his reason had entirely fled, caused by undue excitement about football, which he continually talked about in his ravings. Little hope is held out for his recovery.[9]

Human beings came to seem enormously fragile, prone to excess, and unable to tolerate any level of surprise. There was, apparently, an overwhelming popular refusal to ascribe their fragility to painful social causes or to see society as bigger than the sum of its suffering individuals. Meanwhile these same fragile big-city Americans were expected to carry on at the same rapid pace they had maintained before disaster, both social and personal, struck. The pace of city life was not only a fact but also a genuine point of pride, expressed in the dominant public language of "Keep moving" and "Get a move on, would you?" No one suggested that slowing the pace could be an answer to any problem; in fact, most observers predicted that city pace would continually accelerate and extend its reach into all sectors of life, whether work or play. The going folklore of character suggested that a person who moved briskly and exhibited any small signs of cheer must be doing all right. "The pace that kills," a phrase once applied only to chappies' round-the-clock search for amusement, began to acquire after 1893 a far wider social reach.

Outside and inside, views of the American big-city pace and its effects differed. Foreign observers, who had poured into the United

States throughout the century to stare and analyze and write, *felt* the pace as soon as they arrived in New York City. They not only saw the pace but they also heard its commands and experienced it bodily twenty-four hours a day. They noticed that the pace affected the style and duration of every public interaction, and that "interviews were brief, words few, decisions prompt." In short order, outsiders grasped three driving concepts of American big-city culture: first, that the American city had no moment when the population agreed to stop moving and relax; second, that Americans lived within a general social agreement about the unquestionable value of speed; and third, that the socially agreed-upon public pace was not individually resistible. When M. Bertie-Marriott, *Le Figaro*'s new American correspondent, first arrived in June 1876, he took a boat from Jersey City to New York and felt himself immediately enwrapped in the pace:

> It was nighttime, but nevertheless one felt that these people are always in a hurry, in a hurry to arrive, in a hurry to take their rest, in a hurry to wake up; sleep is wasted time. Today was urging them on toward tomorrow, for tomorrow will be for each of them what yesterday was—a struggle! Time, they say, is money. One must act quickly, for they are all running.[10]

During the 1880s, confident foreign observers fancied themselves making a quick and easy adjustment to the New York pace. In 1882, T. S. Hudson at first found the city's "flash-of-greased-lightning" style to be "very wearing," but announced that "before many hours were over I was hurrying along with the stream." Adjusting to the pace did not, however, mean admiring what the pace had made of one: Hudson managed to simultaneously congratulate himself and deliver an insult when he said that accommodation to the city surround had transformed him into a "bustling, bumptious Yahoo." Similarly, W. G. Marshall, who found New York to be as "compact as a nutshell," claimed that he felt himself "quite at home after a couple of days when in any other place it would take a week." Marshall's complacency, however, leaked away when he took an East River ferry boat to Brooklyn at five in the evening and felt the pace affect his body.

On the arrival of one of these boats there is a regular stam-
pede among the passengers to see who can get ashore first.
Many jump off before the boat has become stationary, and,
like a flock of sheep, those behind them follow suit, and the
deck is cleared of several hundred people—yourself among
the number, for you are irresistibly pressed forward from the
rear and obliged to jump off with the rest.[11]

After 1890, not one foreign traveler had the nerve to claim that he
or she had made an easy adjustment to city pace. The experience of
public transportation—in a time when absolutely everyone used public transportation—stripped them of their confidence and told them
that they were nothing but timid outsiders swept up in a moving mass
of utterly reckless insiders. In 1891, the "rush and turmoil" of the elevated so terrified Alfred Pairpont that he thought he could "scarcely
get in or out of cars with safety before the signal is given and the pant-
ing locomotive is off again." The noise of the elevated terrified Pairpont and the "red-hot cinders" that it dropped to the sidewalks made
him "sick of modern improvements." In 1895, the legal speed limit for
New York City trolley cars was ten miles per hour. Because Americans
found that pace intolerably slow, most trolleys ran at over twenty miles
per hour, and some got up to thirty, a rate of speed that struck foreign
visitors as sufficient to "rival an express train"; they duly reported
numbers of fatal accidents to their readers. They began to admit that
they were too far behind ever to catch up with the American pace, and
that in fact they found the pace intolerable. On his second day in New
York City, Paul Bourget demanded to know:

At what time of day do they die here? At what time do they
love? At what time do they think? I spent a day in the cable
cars, the elevated, the electric cars, the ferry-boats, seeing the
city. One succeeds the other so rapidly, you are so quickly
transferred from tramway to tramway, from train to train, that
the stranger, one who is not up to the times, feels a stupefac-
tion. You miss a car—another is here; so full you could hardly
drop a dollar on the floor. You get in nonetheless, glad of a
chance to stand inside on the platform, on the step, while

· "FARES!"

ragged urchins, frightfully thin, but all nerves and energy, manage to jump aboard of a car between two street corners, into an
elevated car between two stations, crying the daily paper—no,
not even that, the paper of the hour, of the minute.[12]

In 1904, Philip Burne-Jones felt that public transportation was
designed to test his personal fortitude. As a trolley neared him, the
anxious Burne-Jones worked at appearing confident; he was certain
that the motorman studied him from a distance and if he "observed
any signs of weakness or uncertainty," he would not only speed right
on past Burne-Jones but would also shoot him a devastating look of
"contemptuous indifference." Everywhere Burne-Jones went in New
York, locals urged him to "get a move on" and "step lively." Under
pressure, he began to feel "vaguely eager to reach some goal—to
achieve some object," though he had no idea what that object might
be. The American spirit of mobile restlessness, he wrote, was a "contagion" affecting everyone in the city:

The children playing in the streets seem anxiously alert— babies in arms often look thoughtful and careworn, and glance sharply up at you, with fatigued, nervous eyes. The very cats appear *distraites*, and preoccupied, and as though they were late for an appointment. Who would ever stop to say, "Puss! Puss!" to an American cat? And the dogs are dreadfully busy, too.[13]

Chicago, according to foreign visitors, was worse. In 1881, Joseph Hatton was riding a streetcar in Chicago when traffic halted while the drawbridge across the Chicago River opened. Hatton was astounded to see men "leap from the streetcar," race onto the moving bridge, and jump to the other side. There the men waited for the drawbridge to close and the streetcar to cross, whereupon they reboarded it. Foreign visitors to the Chicago Stock Exchange became "horribly nervous" after just a few moments of gazing down on the "wild gesticulations, the hideous shoutings, the continual uproar and confusion" on the floor. Robert Anderton Naylor asked, "How can these men stand such a life of excitement?" Chicago, according to Burne-Jones, had "all of New York's least desirable characteristics exaggerated twofold." The streetcars ran twice as fast, the skyscrapers looked twice as high, and the newspapers seemed to him twice as yellow. In 1906, the *Herald*, which had never ceased loathing Chicago, armed its reporters with stopwatches to go out onto the streets of New York and Chicago and establish, "in a spirit of absolute fairness," which city exhibited "the most hustle." The results were inconclusive, a matter of mere "trifling advantages" awarded to one city or the other. "The problem," announced the *Herald*, "is far more complicated than might be supposed." Real stopwatches, according to Elijah Brown, were unnecessary: each individual "pale, cold, nervous Chicagoan" held in his hand a figurative stopwatch and "listened to its remorseless tick. This is life in Chicago." In 1915, by which year automobiles had joined Chicago's traffic stream of wagons, motorcycles, motor buses, and streetcars, Paul Estournelles de Constant pronounced "the combined effect too much for me. Some of the crossings suggest visions of hell, the impression being strengthened by the flashes and strident squeaks from the trolley cars. And yet there are human beings who, instead of being mere visitors like myself, live here!"[14]

For Chicagoans, it seemed, insanity was next to unavoidable. In 1911, Annette Meakin identified in the very air of Chicago an active element that imbued locals with "intoxicating exhilaration" and drove them to "live a life of feverish activity." Exhilaration was temporary, for in short order the fever, according to Meakin, caused the "health and brain-power" of the individual Chicagoan to undergo "sudden and terrible collapse." Meakin located her proof in the "innumerable sanatoria on the shores of Lake Michigan filled to overflowing with men and women who have thus broken down, and become victims to the worst forms of America's particular complaint—nervous prostration; and in thousands of cases of insanity." Medical experts agreed: the *Herald* gleefully announced that in papers read before the Anthropological Society in Chicago, Dr. Adolph G. Vogeler had reported 85 percent of Chicago's population to be insane, and his colleague Dr. T. B. Wiggin had further proved that the remaining 15 percent were "on the verge of mental breakdown." The two physicians found evidence of lunacy in Chicagoans' excitement over baseball and in their excessive postcard collecting, dancing, strenuous application to business, and "lack of appreciation of poetry." Dr. Wiggin, "in responding to his critics, said that sleep and rest for prolonged periods were the only means by which mental equilibrium could be restored in nervous Chicagoans."[15]

Like any tourists, foreign visitors who took their observations on city streets saw only the "front" of American big-city society. Those who stayed longer and looked harder observed that successful Americans did not "run up and down in a form of busy idleness" but "sat still, surrounded by bells and telephones, and moved only to eat or sleep." Others noticed that the individual big-city American seemed not personally agitated but patient, undiscouraged, firm in the "belief that he always has a chance." Most characteristically he exhibited a "cheerful fatalism that follows him everywhere in his life, like a kind of optimism." Outsiders who were untiringly suspicious, however, glimpsed behind surface optimism an "anarchic exaltation of irresponsibility," and behind Rooseveltian strenuousness a mere national twitchiness. It was the city grid itself, they charged, that created an illusion of speedy human movement, and American success had been achieved through the sacrifice of both manners and attention to the fine arts. They asserted on the one hand that Americans lacked any concept of leisure and on the other hand that Americans cared only for fluffy amusements: "The Americans don't want to THINK." On occasion they were prophetic. In 1908, Charles Whibley, after a close study of "the apparatus of life in New York," its elevators, electric lights, and telephones, located the source of America's public rush in the big-city trade-off between privacy and technology—a trade-off that would increasingly affect Americans throughout the twentieth century.

> To win all the benefits which civilization affords, you must lose peace and you must sacrifice privacy. The many appliances which save our useless time may be enjoyed only by crowds. The citizens of New York travel, live, and talk in public. They have made their choice, and are proud of it.[16]

On the inside, from far behind the public-transportation "front" that terrified visitors, the public pace took on a different and an even darker coloration. Whereas outsiders' descriptions of American big-city life were low on metaphor—aside from their standard "vision of hell"— inside analysts produced thickets of metaphor out of the amusements, technologies, and big consumer icons of American life. The diamond extended further control over American life when big-city women were claimed to be as "cold and glittering" as diamonds, and when city life was said to be so emotionally cold that the occasional rare "bit of sentiment" was like "a diamond found in the dirt." Amid a palpable sense of uncertainty that took hold well before the crash of May 1893, the *Herald* began to represent city life in the win-or-lose extremes of gambling—before shifting into metaphorical weather, religion, undersea monsters, bodily organs, and human emotions.

> After all, this New York existence is much like a roulette wheel. We either play on the lucky red or else we are swamped in the black of despair. It seems that a New Yorker has but two moods—either he is very happy and confident or else very gloomy and despondent. Of course this condition is an unnatural life and a very short-lived one, and yet it is infec-

tious. It is in the air. It's always a glorious sunrise or a clouded midnight to the average man who lives on Manhattan island. The medium-spirited man does not exist. He is of Philadelphia, or, mayhap, Boston, but not of New York. If he chances to come here he is first astonished and then a willing convert. The spirit of unrest is an octopus whose arms are closely hugged about this loose little island of ours. Money is the heart that the monster's tentacles reach for. And money is the heart that we all try to protect from his ravages. The love of it here surpasses that love of scriptural lore.[17]

In metaphor, hard times assaulted the body itself: a given individual might imagine himself to be "a bunch of nerves connected by hairsprings" or "played upon by a thousand wires" that "compelled him to be a sharer in the universal uneasiness." City people pictured themselves as hit by the hammer whose fall signaled the close of the trading day, or as clubbed by "the sledge-hammer of progress." A metaphorical language of emotional extremes and sheer terror appeared in many contexts: individuals *trembled* and *shook as if palsied,* they were *smashed, ruined, jammed, crushed,* and *trampled,* and they *groped for assistance* in the face of a *roaring vortex,* an *onslaught,* and a *general howl.* They *twitched, screamed, cried out,* and *wept.* They made "strange leaps in the dark" and engaged in real or metaphorical somnambulism. Many escaped both reality and metaphor by dropping out of sight without apparent reason. On February 11, 1894, Dr. A. E. Osborne read before the Medico-Legal Society in the Academy of Medicine a paper titled "Strange Disappearances." Considering the numbers of Americans who had vanished in the previous nine months, Dr. Osborne wondered if shape-shifting was possible, if "the human face and form can within a few minutes under proper circumstances assume conditions foreign to the original stage." Americans might thus, suggested Dr. Osborne in an extraordinary turn of phrase, render themselves "easily unrecognizable."[18]

Across the next dozen years, images of money and technology—joined to bodily unease—worked their way ever deeper into city-dwellers' visions of themselves. In a lecture titled "Delusions of the Sane," delivered to the Entertainment Club in the Waldorf on October

20, 1906, Dr. John D. Quackenbos described big-city Americans who had undergone "nervous bankruptcy," who could not bear to shake hands with others, who imagined themselves to be automobiles going at high speed, who thought their human vitality had been altered by the ingestion of "machine-made food," and who spent their days agitatedly "shopping even though they had no money to shop with." Then Quackenbos relayed to his "large and fashionable audience" of intent listeners the story of a "well-known author":

> She believes that she is under the hypnotic power of a physician who sends current and vibrations into her vital parts with the purpose of appropriating her energies for his own use. She feels the clutch of his hand; he comes into her room in the form of a cloud; his astral being flaps its wings beneath her bodice as a spirit bird, which sucks from her soul the life blood of her genius.

Then Dr. Quackenbos delivered up a surprise to his hundreds of "intent listeners": the "well-known author" he had described and every other case he had mentioned *were just as sane* as any other big-city American.

Outsiders prescribing for Americans usually suggested a nice vacation, but close observers of American-style vacations glimpsed in them no relief from pressure: vacationing Americans, just like working Americans, required that "everything be arranged to suit those who have no time to lose." And although an outsider occasionally suggested that Americans might cultivate the inner life, insiders perceived their fellow citizens as "singularly inapt to suffer from such ills as prick the soul," and never took that tack. Insiders looked for relief in neither escape from the city nor interiority but in public life, in consuming not less but more of what the city offered. More shows and funnier shows, suggested the *Herald,* could be curative: a professional entertainer who offered "mirth as a medicine" could single-handedly "lighten the burden of modern society." The entertainer Marshall Wilder explained that "the more intent the mind is, the more diverting and alluring must be the agency that would induce it to pause for a time."

In Wilder's view of the past, the "old fogies of the back centuries were not intense" and thus were too easily amused, whereas "the blasé mind of today needs spice, and the entertainer must be that spice." Wilder's conclusion? "The people want more fun."[19]

The entertainment cure, however, involved neither relaxing nor slowing down; it might well require stepping up the pace and even racing from one curative entertainment to the next. In 1909, John Van Dyke described the New York City postentertainment rush:

> After the play or the show is over all the exits are hastily thrown open. People cannot get away fast enough by the main entrances. They may stop a moment to talk to some acquaintance, but usually they lose patience with anyone who holds up the line of people on the stairways or in the vestibule. Just why or what their haste they scarcely know. Most of them are going home and to bed, and are in no hurry about it if they stopped to think, but they understand they must move at the New York pace or else be stepped upon. Thus it is that the average person gets out of a theater faster than he went in; and after he is alone on the sidewalk he perhaps stops to think what he will do or where he will go.[20]

Once out on the street, already fast-moving theatergoers were literally thrown into their carriages by theater attendants, and sometimes suffered the loss of fingers crushed in carriage doors that attendants slammed shut prematurely. Persons leaving the theater on foot liked to see further little shows performed on the sidewalks. Public fasters were a street vogue in the early 1890s but fell out of favor after the crash: "Americans," said the *Herald*, "want to watch someone doing something, not someone whose specialty is not doing something." Consequently, gross feeding, for example, became a popular sidewalk show; a man who promised to eat a hundred eggs in a hundred minutes could draw a small crowd. Pausing to watch a little sidewalk show, however, meant possibly causing a few seconds' delay to another city person who was in a hurry, and that prospect infuriated the *Chicago Tribune*. In an editorial on June 8, 1893, the *Tribune* announced that the pressures of city life had to be addressed on the sidewalk level. No force could make Chicagoans keep to the right on the sidewalk because, according to the *Tribune,* Chicago held "one hundred and twenty thousand persons who cannot discern their right hands from their left." Therefore enforced constant motion was the answer: "Men and women must be compelled to keep moving. Even if there is something attractive in a shop window people must not stop in front of it."[21]

If a proliferating body of public advice on a subject is any guide to the state of society, then the American emotional state after 1893 was troubled indeed. Much of the advice was couched in money language, thus replicating within the individual the original cause of his or her stress. In *The Efficient Life,* Dr. Luther H. Gulick claimed that the "strenuous life" advocated by Theodore Roosevelt "*taxed* too extremely the *capital* of life force"; Gulick suggested that Americans strive instead toward a new curative goal: *efficiency.* Likewise Horace Fletcher, in *Menticulture, or the ABC of True Living,* attempted to banish "anger and worry" from American life by labeling them "the most *unprofitable* conditions known to man, *thieves* that *steal* precious time and energy from life." The *Herald*'s book reviewer judged Horace Fletcher to be quite wrong. Clinging to that newspaper's long-time belief in the value of delusions, the reviewer rejected the language of finance in favor of the language of progress, and opined that worry and anger were "essential factors in progress, based upon a hope of something better in the future. That something better may be a delusion but the effort to reach it raises the individual by just so much above his former status." Other advice-writers suggested other delusional escapes: American fans of Jane Austen, always a sizable group, were advised to re-create in the American city "the good old traditions and kindly wholesome life of a quiet Anglo-Saxon home." Even the spirit of Henry David Thoreau made an unusual turn-of-the-century reappearance when Bolton Hall, in *Three Acres and Liberty,* advised that men forced "to drop out of the city's ranks by the pressure of an overstrenuous life" might consider "the vast and unknown possibilities of the very small farm intensely cultivated."[22]

A different and far more clamorous advice literature of the 1890s, however, told suffering Americans to look not to their society or their specific living environment but to their own bodies, their state of health and their personal appearance. Although this literature's metaphor was disease, the expert authors did not recognize it as metaphor. New York diet specialist Helen Densmore classified both old age and obesity as real diseases, curable only if the sufferer never again tasted wheat, cereals, bread, or potatoes. The very best diet, according to Densmore, consisted of nothing but fruit and nuts, but late-nineteenth-century America's "digestive organs" were already too "vitiated" to allow them to eat nuts. The obvious nut substitute was meat, and consequently Densmore's complete diet regime incorporated meat, fruit, eggs, milk, and cheese. The popular and influential French diet authority Maurice Verneuil undercut Densmore, however, with his claim that meat-eating—especially if the meat was pork—caused cancer. In 1891, a Dr. Gibbs, who like Densmore defined obesity as "a *disease* easily cured with heroic self-sacrifice and the exercise of the greatest possible will power," rose to temporary prominence. Gibbs had reduced President Grover Cleveland from 307 to 235 pounds through application of the Schweninger cure, a regime which embraced both beef and mutton fat. Two years away from the cure, however, Cleveland was back up to three hundred pounds, and Gibbs was called in again. He told the press that Cleveland's "fat has been accumulating at the rate of more than half a pound a day. It encased him from head to foot, and layer after layer was added with a regularity that became alarming. To get around became an effort for him." Gibbs denied that the original Schweninger cure had failed; he claimed that Cleveland's weight problem had its source in the "peculiarly fattening" climate of Washington itself.[23]

Grover Cleveland's diet doctor was outright opposed to exercise; he assured Americans that a person who exercised actually gained poundage thereby. Rich Americans who tried European weight-loss plans were always told that America's rich food and the "national disinclination to take the requisite amount of healthy exercise" were killing them; at home, however, plenty of physicians and experts agreed with Gibbs on the deleterious effects of exercise—especially on women. Because they saw Americans as prone to excess, they defined American exercise as, by nature, nationally excessive. The fearmongers among them maintained that too many hours on the bicycle produced an ailment called "Bicycle Kidney" and that too much rope-skipping induced peritonitis. In the 1890s, Americans were further overwhelmed by life-threatening excesses not of their own choosing: they hiccuped themselves to death, and they yawned so repeatedly that their jaws went out of joint, and they sneezed so uncontrollably that they blew an eyeball across the room. They were attacked by excess from the outside when, in 1899, monstrous swarms of "new and terrible" insects called "kissing bugs" and "strangling bugs" converged on New York City and frightened the residents. Finally, in 1899, a new disease brought the city pace to a halt: numbers of big-city Americans began to ossify, to turn to stone from the toes up. "Ossification," announced the *Herald,* "is the disease of the moment."[24]

In the face of such terrors, many big-city Americans turned to, in the words of Emily Dickinson, "those little anodynes that deaden pain"—and also to those little stimulants that banish gloom. Aside from arguments over the nature of dependence, American fondness for anodynes and stimulants raised questions of public behavior: who could publicly ingest what? How might signs of ingestion best be handled in a world where everyone was watching, looking, and comparing? The answers to these questions were gender-divided. The dominant urban view of alcohol was simply this: "Men Drink It." In contrast, a woman with a drink in hand, even in total privacy, was represented as a clear danger to society. Women themselves publicly and regularly reinforced the gender split on alcohol consumption. In 1894, for example, the immensely prolific and popular mystical novelist Amelia Barr asserted, "Men may drink with far greater impunity than women, both personally and socially. Personally they have more will power to resist the domination of alcohol and socially their influence and example as slaves to it is far less disintegrating. It is a great truth that vice in a woman is more of a crime than the same vice in a man."[25]

Regardless of such efforts to set rules, both men and women drank and both concealed the conspicuous effects of alcohol with spices. Women denied entry to New York City's ten thousand barrooms treated their pain at home with Jamaica Ginger, a suspension of ginger in alcohol that delivered the stimulant and the spicy cover-up simultaneously. Saloonkeepers set out on the bar little trays of "cloves, pep-percorns, allspice, bits of cinnamon and scraps of lemon peel" for ingestion by male drinkers. A physician described as "a man of the world"—meaning that he also took a drink now and then—claimed that men habitually popped a spice of their choice into the mouth and chewed it in order to "remove the taste of the liquor from the drinker's mouth, or the odor of it from his breath." Prone to excess in spice-chewing as in everything else, such men were quickly "enslaved to the spice habit": they took to carrying pocketfuls of cloves and pepper-corns on which they "nibbled during their working hours." The especially dangerous cloves, explained the doctor, remained in the stomach "unassimilated, serving as a nucleus for the crystallization of alkaline properties, until finally they gather a coating which gives them the appearance of globules of glass."[26]

Slave and *fiend* dominated the public language of anodynes: opium fiend, slave to morphine, slave to liquor. Men attempted to drink out of reach of that language when they called cocktails *things* and *smiles,* but only the stimulant cocaine, newly arrived in American cities around 1895, escaped—for a time—the language of fiendishness and enslavement. Cocaine, according to David Courtwright, emerged as a common therapy in America in the mid-1880s:

> Like morphine (but unlike heroin), cocaine was recom-mended for a wide variety of conditions. As one advertising brochure put it, an "enumeration of the diseases in which coca and cocaine have been found of service would include almost all the maladies that flesh is heir to." Cocaine was rec-ommended as an antispasmodic, aphrodisiac, anodyne, and local anesthetic, as a specific for hay fever and asthma, and as a cure for alcoholism and opiate addiction, to name but a few of its proposed uses. It was also recommended as an all-purpose tonic, for patients who exhibited "melancholy" or "the blues" or other less than precisely defined depressive symptoms. One especially popular product was Vin Mariani [widely advertised in New York City newspapers], a coca wine used and endorsed by Americans of no less stature than Thomas Edison and William McKinley.[27]

Cocaine was further available in soft drinks and in patent medicines—so called even though not one of them was ever patented. Cocaine in pure form could be mail-ordered. Big-city Americans, looking for stimulation and a feeling of increased energy, loved cocaine—but so did small-town Americans, and in late 1896 their love for it became public.

Early in 1895, a millworker from Cheney's Silk Mills in South Manchester, Connecticut, traveled eight miles west to visit Dr. Jonathan S. Curtis in Hartford. The millworker complained to Curtis of catarrh—in other words, he had a persistent head cold. Curtis wrote a prescription for a "volatile snuff" compounded of "sugar of milk, magnesia, menthol and cocaine," and the millworker returned to South Manchester, where he had the prescription put up at T. Weldon & Company's drugstore. Since South Manchester had always been "a great snuff-taking district," the compound had a special local appeal, and the millworker found it "such an effective and pleasant remedy for colds and kindred ailments" and "so exhilarating" that he passed on the prescription to coworkers, who wrote out their own copies of it and had the prescription filled. The snuff was especially appreciated in the velvet rooms at Cheney's Mills, where the lint was thickest and breathing the most difficult. Copies of the prescription traveled through every department of the mill, and spread from there throughout the town of South Manchester. By the fall of 1896 the cocaine snuff prescription had been copied and refilled perhaps a thousand times over.

Eventually every South Manchester local—men, women, girls, boys, the very old and the very young—was buying cocaine snuff in ten-cent packets or twenty-five-cent vials. At the local drugstores, one of which was operated by a "leading doctor" in town and another by Cheney the mill owner himself, druggists ceased requiring a look at the much-copied prescription; they prepared the compound in "big lots, so as to have it ready for the frequent demands" and they kept a big jar of it right to hand on the counter. In fact, although the druggists claimed to "view the growth of the habit with alarm, they were coining money out of it." A few local doctors tried without success to counter-prescribe, but most of South Manchester's doctors saw their caseload decrease; they guessed that they were being avoided by longtime patients who "feared that the stuff would be taken away from them" by the doctor. The doctors themselves admitted that they had no idea "just what the result of deprivation might be."[28]

On December 29, 1896, a *Herald* reporter went up to South Manchester and described his entry into the town.

> A stranger going to the pretty little place on an electric car will first see the conductor or motorman take out a bottle, shake a white powder in his palm and then snuff it with intense satisfaction and a long-drawn sigh of relief. In the street every other man or boy is using the stuff. Strangers stop each other at the corners and ask:—"Can you give me a pinch?" The question is readily understood and the bottle of snuff is forthcoming.

Hearst's *Journal,* an operation that frequently discovered what was news by reading the *Herald,* thereupon rushed its own reporter to South Manchester to visit local drugstores.

> The reporter visited a number of drug stores, the first one being Horton's, and asked for a preparation of cocaine and menthol, "which is good for catarrhal troubles." The clerk at this store refused to sell the snuff without a doctor's prescription. At the drug store of Dr. Weldon, one of the prominent physicians of the town, the reporter bought some cocaine snuff without any questions being asked. "How much do you want?" said the clerk, mechanically, taking a spoonful of the white powder from a jar which stood conveniently near the counter, and which was at the time nearly emptied. "Half an ounce," was the reply. The charge was 25 cents. Entering the next shop—Belcher's—the reporter said, carelessly: "Ten cents' worth, please, of cocaine snuff." It was given at once, without a word. At Cheney's, the next drug store, there was the same success.

Studying South Manchester residents on the streets, both the *Herald* and *Journal* reporters were struck by their watery bloodshot eyes and by a universal pallor that local informants laid to their inability to sleep. A drugstore clerk told the *Herald* that locals "come in here at all hours of the day and say, 'Give me some of that snuff, quick; got to have it.' But that isn't the worst of it. They come to my house after I am asleep and make me get up, go to the store and satisfy their cravings. They cannot sleep and they don't know what they are about much of the time." Another druggist admitted "making a handsome profit from the drug," but claimed he would "cease to handle it after the present supply is exhausted."

At some point during the two years that it took the town of South Manchester to become fully habituated to cocaine, a Cheney's Silk Mill worker—said to be suffering from catarrh and thus equipped with the snuff prescription—moved to New York City, found work, and told her coworkers about the prescription. Anodynes and stimulants were of course well established on the cityscape; the year 1897 marked a heyday of drug use and a national high point of addiction. As the *Herald* liked to put it, many New Yorkers, both men and women, were "slaves of the poppy's spell" and frequently "saw the china pug wink." The stimulating cocaine snuff, however, took an interesting path upward from factory workers to stage people to the rich, and its use became more private at each step on the social ladder. In South Manchester, reporters had seen none but boys and men publicly ingest the snuff, but in New York City snuff use was rewritten as a gendered habit: news stories protected men and insisted that only women used the snuff, and at that only in private. Evidence of the compound's arrival in New York City, in fact, consisted solely of a steep increase in the separate sale of each of its four ingredients. Persons in charge at the wholesale and retail drug houses in the city guessed, for reporters' benefit, that "many women, in order to hide the fact that they are using it, prefer to purchase the ingredients separately, and often in different drug stores," before making up their own snuff at home. The manager of Schieffelin & Co. told the *Journal* that "magnesia and sugar of milk when powdered with cocaine crystals serve to give body to the mixture, and the menthol or peppermint extract gives flavor to it. I have heard of the craze for cocaine snuff that has sprung up," he said, "and was not altogether surprised, for with the extensive use into which cocaine has come and the unrestricted way in which it was handled it was almost certain to be taken up as a stimulant as soon as the people learned about it."

Anodynes at the end of the nineteenth century, as has been frequently noted, were either far too powerful for most complaints or far too weak to relieve any complaint. Some persons of means, for example, who had been overstimulated into "nervous prostration" were treated with "a subcutaneous injection of phosphate of soda," claimed to "act directly on the nervous system, a portion of the organism for which it has the greatest affinity." Experts said the treatment was "utterly useless." Rich Americans who had become dependent on alcohol, morphine, and cocaine checked into one of the Keeley Institutes to undergo the Gold Cure. Besides lounging on the veranda and playing tennis and croquet, these patients took a hypodermic injection of "a pink fluid" four times a day and swallowed a teaspoon of Leslie Keeley's mysterious "bi-chloride of gold" every two hours. Critics and rivals asserted that Keeley did no more than hypnotize his clients into believing themselves "cured," but in the mid-1890s only a "Diamond Cure" could have rivaled the Gold Cure for sheer iconic name appeal. The beautiful notion that gold—aside from all its other wonderful uses—had actual curative powers was great enough to keep Keeley afloat well into the twentieth century.[29]

To many observers both inside and outside American society, the 1890s looked like a heyday for "anodynes" in the United States. Outside observers who hated everything about America—British writers on the American scene such as Patrick Vaile, Roger Pocock, and Rudyard Kipling—looked away from the intense social pressures put on Americans by business failure, the high-pressure nature of work for those who had work, the hopelessness of being out of work, and the continuing demand that everyone move faster. They instead seized the occasion to picture America as a nation of addicts and habitués, incurables suffering from a great national character flaw. Roger Pocock's 1896 diatribe in *Rottenness* is typical of such views:

> The American citizen is nothing if not reckless. The American diet of iced drinks, iced confections, candy, tea, coffee, pie, hot cakes, and other slow poisons, aggravated by haste and vorac-

ity, has made the nation dyspeptic. The druggist has an illicit bar in a back room for sneaks, but even the accommodation of secret drunkards is not the worst phase of his business. Want of sleep, neuralgia, and tremulous hands are indications of nerves jarred by an unhealthy life. But instead of mending the life, the fool soothes the nerves. Opium, laudanum, morphine, chlorodyne, chloral, chloroform, hashish, quinine, anti-pyrin, anti-fibrin, bromide, cocaine, the injection under the skin, the sugared pilule, the drops from a phial, the tiny opium cigarette—all these seductive measures beguile them to sleep, relieve pain, or tranquilize the nerves, and make them ask for more and more and more. The skin of the stockbroker is pimpled with marks of the syringe, the fashionable woman *whispers* her orders across the counter. One sees the druggist's victims among the human elements in the streets, listless, haggard, sallow, with haunted eyes, crawling languidly to their death from heart failure or from sheer incompetence to make a living. And these are becoming more numerous than the drunkards.

No inside observers engaged in such unsympathetic and sweeping condemnation of Americans, but plenty of insiders were no better at social thinking than was Roger Pocock. As social misery spread, Americans who saw no way out of failure did not cease requiring themselves to succeed. Not even in death could they imagine an end to their struggles; they envisioned themselves trapped in a city that was a "vision of hell" wherein they were condemned to "live upon the earth forever, suffering eternally the pangs of nervous exhaustion." Furthermore, they blamed only themselves for their predicament. Some, in the twilight of a century that seemed to be ending so badly, dealt with their situation by applying increased pressure not only to other humans but to any living being unfortunate enough to stumble into their circle of power.[30]

NOTES

1. *New York Times* and *New York Herald,* 17–18 November 1893.
2. *New York Herald,* 18 November 1893.
3. *New York Herald,* 1 May 1894.

4. *New York Herald,* 13 February 1894.

5. *New York Herald,* 4 November 1894.

6. *New York Herald,* 17 February 1894.

7. Rosenberg, 8. A few foreign visitors glimpsed the confinement of those judged insane. Philip Burne-Jones, for example, wrote of taking "a little expedition one day on a government tug, which took one past the simple houses of old New York, among innumerable islands, nearly all of which seemed to be dedicated to refuges of various forms of sin or suffering. Every time I asked what a building was, they told me, 'Oh, that's a lunatic asylum,' or 'That's such and such a prison'; and far off I could see mournful little processions of afflicted individuals waving handkerchiefs to us as we passed" (100–101). In 1894, the *Herald* began chasing numerous stories of persons wrongfully confined, and in October found Ester Hleis confined in a city asylum. A Syrian immigrant, Hleis had no English. She had "gone out with a little girl to peddle some Eastern wares in the street. They became separated. The woman could not find her way home, and after wandering for hours was discovered crying by a policeman who promptly took her to a station." Because the terrified Hleis continued to cry and to kiss a crucifix she was carrying, she was bundled off to undergo a "mental examination." The doctor who examined her believed her to be Italian, but because she failed to respond when addressed in Italian, he pronounced her "a maniac." Within the asylums themselves, doctors sometimes used other patients as interpreters, but they could not appear in court or at inquiries, and "the doctors say their word should not be taken because they are insane. In the asylums, however, they are considered capable of describing the ailments of fellow victims" (*New York Herald,* 14 October 1894).

8. *New York Herald,* 2 March 1893, 15 May 1893.

9. *New York Herald,* 29 August 1899, 17 December 1899, 1 April 1893, 13 February 1893, 19 November 1893, 20 November 1893, 11 December 1893, 7 January 1894, 10 April 1894, 17 November 1893, 25 February 1894, 4 December 1893.

10. Leng, 218–219; Bertie-Marriott quoted in Offenbach, 158.

11. Hudson, 34; Marshall, 22–23.

12. Pairpoint, 151; Bourget, 27.

13. *New York Herald,* 8 March 1895.

14. Hatton, 39–40; Naylor, 95; Burne-Jones, 20–23; *New York Herald,* 21 October 1906; Elijah Brown, 94–95; Estournelles de Constant, 237–238.

15. *New York Herald,* 16 October 1906. Meakin's own statistics on Chicago lunacy are difficult to follow: "To cross the busiest streets you must take your life in your hands . . . add to all this the deafening and never-ceasing noise, and you will not be surprised to learn that insanity is terribly on the

increase in Chicago, and in this favored state of Illinois. In 1908 the statistics showed that one out of every 465 persons was insane. The population has increased 89% in thirty years, and the number of insane persons cared for by the state has increased 369%" (110).

16. Whibley, 17–18; Soissons, 33–35; De Bary, 248–249; Samuel Smith, 19; Money, 227; Felix Klein, 244–246; De Rousiers, 414; Vay de Vaya, 308; Humphreys, 73; Whibley, 22–23.

17. *New York Herald,* 7 October 1894, 29 January 1893. In *Who Tells Me True,* Michael Strange used the same roulette metaphor in relation to her childhood perceptions of her father's life on Wall Street: "Thinking back on those years of my childhood, I can understand a little the strain and exhaustion and final indifference that came over my father's spirit regarding life on the Stock Exchange, if there is such a thing. It became increasingly impossible for him to hold to his path in the vast jungle that big business was causing to grow up. This called for a technique that he had not, nor very likely any of his forebears. He had gotten his living, as they had, from the start of a small inheritance and a decent use of ability. . . . Bankruptcies became more frequent. Incredible appearances and disappearances began to be made through the wheel of fortune. . . . The tempo in Wall Street began to resemble that of a roulette wheel; there was an influx of millionaires and their wives, who bid for their inclusion in society with a lavish display in everyday living, and Haroun-al-Raschid entertainments" (42–43).

18. *New York Herald,* 15 October 1894; Warner on "The American Newspaper" in *Fashions in Literature,* 57; Sherwood, 181; *New York Herald,* 6 May 1893, 12 February 1894.

19. De Rousiers, 308; F. Marion Crawford, *Dr. Claudius; New York Herald,* 8 January 1893.

20. Van Dyke, 214–215

21. An example of posttheater violence from the *New York Herald,* 15 October 1893: "Dr. Fleming May Lose His Thumb. Crushed in the door of his carriage by the Lyceum theater attendant, as a final act of violence. The Caller had thrown two ladies in the vehicle before doing serious injury. Mr. Charles Frohman says his employe was making haste to enable the crowd to get out of the rain."

22. *New York Herald,* 9 March 1907. Significant late-nineteenth-century American writers considered delusion to be a potential though risky answer to human misery. Both Stephen Crane in such short stories as "The Blue Hotel" and Edwin Arlington Robinson in his mid-1890s poems represented human delusion as simultaneously painful and nourishing to the individual; in neither writer's work does the puncturing of another person's delusion result in any-

thing but catastrophe. Theodore Roosevelt, interestingly enough, was a notable public fan of Robinson's work.

23. *New York Herald,* 10 February 1894, 30 July 1893, 19 June 1893, 9 July 1893.

24. *New York Herald,* 15 December 1895, 7 July 1899, 20 April 1894, 14 July 1893, 8 March 1894, 25 February 1894, 27 January 1895, 13 February 1895, 2 July 1899, 3 July 1899, 4 July 1899, 10 July 1899, 12 July 1899, 19 July 1899, 23 July 1899, 31 July 1899.

25. *New York Herald,* 14 January 1894.

26. From the *New York Herald,* 6 May 1894: "Physicians declare that the habit of drinking Jamaica Ginger in large quantities is by no means extraordinary. It is well known that after habitual drunkards have ceased to obtain the stimulus from the ordinary liquors they often select crude alcohol with which to accelerate the heart action. In many instances wood alcohol has been used. Women, as a rule, under these exceptional conditions, select preparations like Jamaica Ginger, which contains some substance which complicates the action of the alcoholic poison and also acts directly upon the digestive organs. Instead of such cases being rare, they are comparatively common, the only difference between the action of the crude alcohol and combinations like ginger and calisaya being that the latter is much more easily treated."

27. Courtwright, 96–98. In 1907, 1908, 1910, and 1913, New York passed a series of laws restricting the availability of cocaine. The 1913 law "placed so many elaborate restrictions on cocaine that legal distribution was practically impossible," 98.

28. All South Manchester material is drawn from the *New York Herald,* 29 December 1896, and the *New York Journal,* 3 January 1897.

29. *New York Herald,* 2 January 1893, 5 January 1893, 1 July 1894. According to David Courtwright, "Leslie E. Keeley was one of the best-known of the nineteenth-century 'cure doctors.' Keeley postulated that inebriety was a disease, and he treated both alcoholics and opiate addicts in his national chain of Keeley Institutes. He also believed that individuals inherited varying degrees of nervous susceptibility, and that opium and alcohol changed or 'educated' impressionable nerve cells. What set Keeley apart was his advocacy of a mysterious specific, the Bichloride of Gold formula, for treatment of both species of inebriety. This opened him to charges of quackery and generated a long controversy" (215–216). Keeley, according to H. Wayne Morgan's *Drugs in America,* "wrote some 500 doctors for their opinion of gold in therapy and began experimenting with various compounds in the late 1870s." In marketing his bichloride of gold remedy, he "secured patents and trademarks for the

bottle's shape, the label designs, and other accoutrements but never sought protection for the formula, in order to avoid stating its composition. At no time did Keeley or anyone else reveal the formula, and it died with the partners" (77).

30. Pocock, 86–87; *New York Herald,* 20 October 1906.

WOULDN'T YOU WANT TO SIT ON SOME
ONE IF YOU WERE CHAINED LIKE THIS?

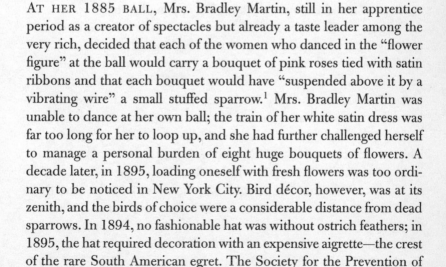

CHAPTER 7

Death in Public

As the returning traveller leaves the shores of America, he remembers most vividly that he is saying good-bye to the oldest land on earth. America is not, never was, young.
—Charles Whibley, *American Sketches*, 1908

AT HER 1885 BALL, Mrs. Bradley Martin, still in her apprentice period as a creator of spectacles but already a taste leader among the very rich, decided that each of the women who danced in the "flower figure" at the ball would carry a bouquet of pink roses tied with satin ribbons and that each bouquet would have "suspended above it by a vibrating wire" a small stuffed sparrow.[1] Mrs. Bradley Martin was unable to dance at her own ball; the train of her white satin dress was far too long for her to loop up, and she had further challenged herself to manage a personal burden of eight huge bouquets of flowers. A decade later, in 1895, loading oneself with fresh flowers was too ordinary to be noticed in New York City. Bird décor, however, was at its zenith, and the birds of choice were a considerable distance from dead sparrows. In 1894, no fashionable hat was without ostrich feathers; in 1895, the hat required decoration with an expensive aigrette—the crest of the rare South American egret. The Society for the Prevention of Cruelty to Animals cried out against the slaughter of egrets but could

not catch the ear of the fashionable. The *Herald* believed the egret battle to be hopelessly lost: "Besides the members of the league formed
for the protection of birds, how many can think of the dozens of egrets
killed simply for their plumage when the feathery tufts of the aigrette
surmount such pretty faces? Very few, I fear. It is ever the gay birds
that suffer."

The other "gay bird" of the moment was the hummingbird. On the
street in 1895, an up-to-the-moment woman might wear a black chiffon neck ruche, "fluffy and billowy around the throat. In the back,
with its wings and tail outstretched and its bill lost in the dusky tissue
was a little dead hummingbird." For frontal bird display, another
ruche, this one of wide green ribbon edged with white lace, held two
dead hummingbirds, "one on either side in the flutes that flare out
under the ears." Even the *Herald*'s dedicated fashion writer, however,
found this "latest modish whimwham" indefensible, and judged dead
hummingbirds disposed on a chiffon ruche to be "ghastly, scarcely
artistic or beautiful." As always at the turn of the last century, when *art*
failed as a cover for questionable human behaviors, matters were
about to get serious. The question at hand was human use of birds
and animals—dead or alive, as décor, as window displays, as pets, as
workers, as menagerie prisoners—in big-city public life.[2]

City people liked to look at animals; they liked European cattle
paintings, Barye bronzes of animals struggling with each other, and
the Central Park menagerie. And even though desire for the ongoing
presence of a live animal was difficult to fulfill in the city, locals kept
trying. Notwithstanding the hard times of December 1893, dealers in
live animals as Christmas gifts did brisk business in birds, dogs, cats,
rabbits, guinea pigs, monkeys, anteaters, honey bears, tame rats, and
white mice. Most popular were small tweeting birds: a single pet shop
in Manhattan in 1893 imported a stock of fifteen thousand canaries
for Christmas, hoping that half of them would survive to be sold. Buyers fought to the death over the best birds. On July 26, 1893, Matthew
Green, a foreman in the street cleaning department, had just selected a
new canary in John Keif's store at 500 First Avenue when he noticed
that James Halstead, proprietor of the Manhattan Dye Works at 130th

Street and Twelfth Avenue, was eyeing the same little bird. Halstead was very drunk.

> "I said that it was a nice bird, and Halstead said that it was not. Then I said that I had heard it sing and it was a good bird. 'You are a damned liar,' he said, and with that he hit me in the eye. I grasped him by the whiskers and shouted to Keif to help me hold him, for I did not want to fight. Instead of helping me, Keif ordered both of us out of the store. We went outside and the fight would have ended right there, but that he called me foul names and finally I struck him, knocking him down." An ambulance doctor from Bellevue Hospital found that Halstead was dead.[3]

Because street vendors sold canaries even in the coldest weather, many buyers assumed that the little birds must be quite hardy, but they learned different once they got them home. In 1893, a letter to the *Herald* complained of the poor condition of street canaries:

> Canaries die or get sick just as soon as you have paid for them. Whenever I buy a canary it seems to be a bird that is especially subject to colds and pneumonia, and it is only by the exercise of the greatest care that I can keep it from succumbing to some pulmonary trouble. Yet the canary bird sellers have their wares for sale in the streets in the coldest weather almost unprotected from the wind. They stand around with them for hours and no bad result seems to come of it. How is it? I give it up. There must be some conspiracy between the dealers and the birds by which the latter die as soon as they are bought, compelling the purchasers to invest in more canaries. You wouldn't think to look at the little yellow fellows that they were capable of so much treachery. Dealers bring them over from Europe with very few precautions against disease or accident, but if I leave one of my canaries alone for ten minutes at a time he swallows a piece of rag and chokes to death. I suppose the whole secret of the thing consists in knowing what you're about, and I don't, at least as far as the canary birds are concerned.[4]

In fact, canaries were disposable. At the 1895 Arion Ball in Madison Square Garden, dozens of canaries and pigeons were released; after the ball, they flew about the rafters until they died. In the mid-1890s, animal buyers further experimented with a variety of disposable reptiles; most were so small and obviously short-lived that they were marketed not as pets but as accessories. In 1893, the chameleon—with a gold chain around its neck—briefly rivaled canaries in the disposable-pet category, but chameleons tended to die quite horribly: seeking to do what they did best, they crawled onto lampshades and "slowly roasted to death while endeavoring to turn the color of the shade—a bilious yellow." Consequently, in 1894, the lizard bested the chameleon in popularity. Women wore small live lizards chained to their belt buckles, and chained lizards were used to enliven the window displays of men's furnishings stores. Although some observers thought it "downright cruel" to immobilize lizards, the little reptiles—who did nothing especially interesting to humans—went out of style before anyone could effectively protest their living conditions. Across just a few months, lizard prices plummeted from a dollar apiece to forty-eight cents to a quarter; then the lizards were "thrown away," superseded by the frogs that became popular in 1896.[5]

In the city, it was finally easier to enjoy looking at an unquestionably dead animal than at a momentarily living one, and the popular furs of the 1890s catered to this desire. Although fur-bearing animals in general had been hunted so persistently by the mid-1890s that some feared extinction of many species, the dominant idea of the time was to wear as much of the animal as possible in as lifelike a form as possible. In 1895, Bloomingdale's advertised "ten thousand fur neck scarfs, with animal heads and open mouths," while C. C. Shayne at 124 West Forty-second Street offered "collarettes and boas trimmed with tails and heads, some of which have rhinestones for eyes and are a great bargain for those who are ultra-fashionable." By 1896, everyone, according to the *Herald*, wanted whole-animal display:

> This winter's garments are to be trimmed with tails, heads and paws, and the fashionable fur this year is grebe. Ermine's day is passing, and though it is still used a little, grebe far sur-

passes it now in popular favor, made up in capes, collarettes, muffs, everything except a whole jacket, and always expensive. The handsomest capes are made with a yoke of some other fur, Hudson's bay otter, sealskin or Persian lamb. Around the yoke there is a trimming of little heads of the same fur, with their tiny eyes shining. Animals' heads, tails and paws are used on almost all fur garments. Jackets this year are trimmed with great bunches of tails fastened on with a large head. Very handsome sable capes are made of the whole skin of the animal, head, tail, paws and all, the tail and hind paws making a fringe around the bottom of the cape, the head and front paws running up on the collars, just like a row of little animals.[6]

Fur could be made up into a complete illusion of life: with a "muff made of a whole fox skin, the hind paws and tail on one side, and the head and front paws on the other," a woman could look "as if she was carrying around a little pet fox or pet dog." Dying animals were equally interesting. Although the bear was the most "popular" animal of the 1890s, Philip Leibinger of Brooklyn, who was "fond of shooting wounded grizzlies with a camera," marketed in 1894 a Christmas gift book of "photographs of bears he had shot and then photographed while they were fighting for life against his pack of Irish terriers, who love nothing so well as dodging the ferocious slaps from the paws of a dying bear."[7]

Disease and death were part of the public life of working animals and of common cats and dogs. No one claimed to enjoy the sight of working animals struggling and dying on the streets, but city people were accustomed to that phenomenon. In the terrible winter of 1894 alone, twenty-five thousand horses were injured or disabled in the "briny slush" created by salt sprinkled on the car tracks. In the 1880s, a city dog's life was also seasonally at risk. In the summer months there were no dogs to be seen on the streets, as William Hardman remarked in 1883.

Among things conspicuous by their absence in the streets of New York are dogs. The visitor to that city will, after a day or two, begin to wonder, and to ask the question, "Where are

the dogs?" Beyond a few very small pet dogs carried by ladies
there are practically no dogs to be seen in New York after the
1st of June until the summer is past. Whether canine rabies is
more common in summer than winter is a question open to
doubt; nonetheless every year at this season the mayor
appoints the official dog-catchers, pound master, and other
necessary persons to carry out the city regulations. Every dog
not properly muzzled and led by a string is captured and
taken to the pound, a temporary structure erected annually
on the most convenient available site on the bank of the East
river. Once there, if not redeemed within twenty-four hours
by payment of twice the amount the city pays the dogcatcher, the dog's fate is sealed. He is put, with other companions in a similar predicament, into an iron cage, which is
swung out over the water and then lowered into it until all its
occupants are dead.[8]

A decade later there were no more dogcatcher patronage appointments: the Society for the Prevention of Cruelty to Animals seized legal
authority over New York City's animal population and sent out its agents
"in full uniform" to handle all dog-catching. The society was equipped
with patrol buggies, electric alarm systems, its own room in the new
Criminal Court Building, and "airtight chambers" for gassing small animals that had previously been drowned. The SPCA thought of itself as
"forming a complete chain of vigilant benevolence," and it added further
links to that chain in 1894 when it took over dog licensing:

Under the old law the Mayor's Marshal issued licenses for
dogs at $3 each, and every dog had to wear a tin-cut tag and
go in leading strings or wear a muzzle. Now for $2 the little
dog may roam at his own sweet will without leash or muzzle if
he only wears "at all times a collar about his neck with a
metal tag attached thereto bearing the number of his license
stamped thereon," which is provided by the society without
extra charge.[9]

In the 1880s, cats were briefly made into clothing: a guest at the
1883 Vanderbilt fancy-dress ball appeared in a costume whose skirt

was a stitched-up display of the heads and tails of dozens of white cats. Manhattan's cat population was immense, estimated at four hundred thousand in the early 1890s. The public life of city cats took a drastic turn toward the uncertain when, on the morning of September 12, 1893, thirteen dead cats were found lying close together in Grove Street Park, and several more lay dead in front of 176 Waverly Place. Groups of three and four cats, pads skyward, appeared here and there on the streets of the neighborhood. Residents "picked their way among the slain" and told stories of hearing, late at night, "the plaintive wailing of an unfortunate tabby cease with a suddenness suggestive of untimely taking off." Street urchins reported seeing in the neighborhood "a mysterious woman" wearing a gold star-shaped badge and carrying a large basket. The woman reportedly used catnip as her opening lure, further gained the cat's confidence by offering bits of meat or fish, and then swiftly transferred the cat to the interior of her basket and disappeared. Postcards dropped overnight on Waverly Place doorsteps requested that residents send "all information about cats in need of food, medicine, cheerful companionship or relief from earthly pain" to 1397 Broadway—a bookstore that functioned as a mail drop.[10]

Over the next several days the newspapers learned that the mysterious woman with the basket was a member of the Midnight Band of Mercy, and that the Midnight Band was a branch of the Henry Bergh Circle of the Order of King's Daughters, a charitable organization whose members "selected their own particular lines for doing good."[11] No names of the twenty Midnight Merciful were forthcoming, but both the Broadway bookstore and the King's Daughters headquarters on West Twenty-third Street were besieged with requests for information about how to be "cruelly kind to felines." Late in September, rumor had it that the Midnight Band was in nightly operation on West Thirtieth Street, near the Nineteenth Precinct police station. A *Herald* reporter lurked throughout the night on West Thirtieth between Sixth and Seventh Avenues, a block that "more cats made their headquarters than any other in the city," until finally he spotted a member of the Midnight Band.

The woman was about forty years old. On her breast she wore a badge in the shape of a gold star, in the center of which was stamped "The Midnight Band of Mercy." She carried a large market basket in which was a dead cat. When she lifted the lid of the basket, which was padded, the odor of chloroform escaped in such strength as to almost overcome one. The cat in the basket was a beautiful black and white animal. He looked fat and well fed. Around his neck was a red ribbon. Alongside him was a big bottle of chloroform.

The woman explained to the reporter that the Midnight Band "did its work of slaying cats for pure love of the dear things" and left the corpses in the street so "the Police could see them and have them taken away by the Board of Health." Although the woman had been "at work" for eighteen hours straight, she said she was "not a bit tired" and had plenty more still to accomplish. The interview grew hostile when it dawned on the reporter that the Midnight Band selected a cat for the chloroform basket not because of disease or disability but simply because the feline was out on the street past eight in the evening, an hour after which the Midnight Band believed a cat had "no business in the street." The Midnight Band member vigorously asserted her right to kill cats: her "work," she pointed out, was no random individual activity but part of an orderly group effort "directed by our president." Then under pressure to cite some authority, she let go of two names: Mrs. David and Mrs. Edwards.

On October 4, Policeman Joe Connelly of the Manhattanville station observed a woman on West 135th Street, near Eighth Avenue, calling to a large tom. She held in her hand a white rag. When the cat approached her, she seized him by the scruff and held the rag over his head. He "squirmed once or twice," but by the time Connelly reached her, the tom was "on his way to cat paradise." The woman "obligingly opened the basket she carried, which was three feet long, two feet deep and a foot and a half across, and Connelly saw it contained six other cats, while he smelt a strong odor of chloroform." Connelly decided to take the woman to the Manhattanville station, where she identified herself as Sarah J. Edwards, a thirty-seven-year-old widow living at 212 West Thirty-second Street. Sergeant Fress quizzed Edwards about what right she had to kill cats.

> "I found the cats out after eight o'clock last evening, and they haven't any right to be," she replied. She smiled when the sergeant explained it was a punishable offense to carry dead animals through the city. He tried to show her the sections in the Penal Code under which she could be held a prisoner, but Mrs. Edwards shoved it haughtily aside and said, "I've killed five thousand cats during three years. I know my business." "I know mine, too," replied the sergeant, and had her taken to the East 126th Street station, which has a matron.

Sarah Edwards spent the night in a cell. The next day in Harlem Police Court, Mrs. David, president of the Midnight Band of Mercy, appeared to defend Sarah Edwards against charges of cruelty to animals. Mrs. David said she had "scientific testimony" to support her methods, she claimed to have authorization from the Board of Health, and she dismissed as a mere wage-earner the SPCA agent who had appeared to support the complaint. Then she played her trump card. She announced that the very next day she would stage a convincing demonstration by chloroforming a live healthy cat right there in the courtroom. Police Justice Welde was having none of it, but he did adjourn the case until the Midnight Band could produce testimony and witnesses in support of their claimed legal authority to kill cats. Over the next few weeks, Justice Welde was bombarded with letters both pro and con, as was the *Herald.* One letter charged that the Midnight Band's real motive was money: "a catgut, divided into sixty-one threads by the professional violin string maker, furnishes 4 E strings at 30 cents each, 15 D strings at 40 cents, and 7 A strings at 25 cents each," for a total market value of $8.95 (equivalent to about $150). The letter-writer's case against the Midnight Band was, however, undermined by his failure to explain why they left their potential profit sources piled up stiff and dead in the streets.

Meanwhile Sarah Edwards herself moved on to Asbury Park, New Jersey. There she wore a conspicuous SPCA badge (an interesting touch, considering that the SPCA did not allow women to join) and carried two baskets—one filled with catnip, meats, bread, and cheese, the other a tin-lined contraption designed to hold a half-gallon jar of chloroform and up to six dead kitties. The Asbury Park Police Department announced that it supported her efforts to reduce the feline population of Asbury Park.

On November 15, Sarah Edwards was tried in the Court of Special Sessions on the charge of "unlawfully, unjustifiably and wilfully killing five dumb animals—to wit, cats." Mrs. Edwards had got herself a new lawyer—none other than Abe Hummel, of Howe & Hummel. Clearly Hummel saw the case as an instructional opportunity: he was "flanked by half a dozen younger representatives of the firm" who observed closely the tactics he employed in "defending the right of a free born citizen to kill cats." Former judge Horace Russell, who appeared for the prosecution, represented the SPCA's interests in maintaining total control over the lives of New York City's animal population. The case was, according to the *Herald,* "replete with human and feline interest." Hummel created a controversy over the color of the cat Mrs. Edwards had been chloroforming when Policeman Connelly apprehended her, and another controversy over the contents of the cat's stomach, and a third controversy over whether "good American cats" would eat potatoes unless at the point of starvation. Hummel produced statistics proving that one thousand kittens were born every day in New York City while the Midnight Band "averaged barely fifteen dead cats in twenty-four hours." He dwelt at length on the humanitarian nature of the Midnight Band and on the simple beauty of their feeding a cat before chloroforming it "so it could die on a full stomach." He suggested that they generally tried, as best they could, "not to kill cats that are household pets." Nonetheless, Sarah Edwards was found guilty and fined two dollars for each "cat murder"; she announced that on principle she would appeal to a higher court.

In May 1894, the SPCA acquired full legal control over cats' lives, though it never attempted to license them as it did dogs. The law effectively ended the Midnight Band's humanitarian efforts in New York City.

Under the new law, the cat, which hitherto has been regarded as a sort of social pariah, has become canonized as it were, and is now recognized as one entitled to the same constitutional rights as her canine enemy and even greater privileges. The cat may revel in front area or back yard, on balcony or fence, without fear or reproach (according to law) and without a license if he or she only wears a collar or ribbon about his or her neck inscribed with the owner's name and place of residence; but otherwise he or she may be "seized and disposed of in like manner as prescribed for dogs."[12]

In the closing decade of the century, city dwellers sensitive to human treatment of animals turned their attention toward the Central Park menagerie, then lodged in nine acres west of the Arsenal. The menagerie drew millions of visitors each year with a display of animals that could be seen nowhere else in the city, and as Roy Rosenzweig points out, "it was one of the few free attractions to which working-class families could bring their children." Nonetheless, across its nineteenth-century history as a public amusement, both locals and foreign visitors expressed discomfort with what they saw there. In 1877, the lion, tiger, and leopard struck William Fraser Rae as "exceedingly unhappy. They are shown not in dens where they may retire from public gaze, and take refuge from affectionate demonstrations conveyed by the points of umbrellas and sticks, but in cages barred all round and placed so that they can be approached on all sides." Some visitors thought the menagerie animals looked angry and terrified; they complained that the animals were allowed to fight with each other, that they exhibited the effects of disease and untended injuries, and that they looked generally "shopworn"—a true consumer's judgment. For others, however, the animals' drawing power lay in their suffering: Blanche Oelrichs, who was not allowed to visit the menagerie because her mother had "fancifully concluded that germs were more prosperous in parks than in streets," had nonetheless heard so much about the misery of the animals that she longed for "just a glimpse of the polar bear which was described to me as having taken his captivity the worst of all."[13]

In May of 1893, concerned persons, charging that the animals were neglected and many reforms were needed, launched a "crusade" against the Central Park zoo. On June 5, 1893—a summer Sunday when the Seventh Regiment Band gave a free concert and fully one hundred thousand persons flocked to the park—the *Times* dispatched a reporter to make a "personal investigation" of the condition of Central Park's goats.

> The Mall was thronged, and children were present in great numbers. Of course they all wanted a ride in the goat carriages, and the goats, who are worked from eight in the morning to sundown, seemed exhausted by their labors. They were kept continuously on a trot. The carriages weigh not less than 100 pounds, and each accommodates six children. Sometimes seven were crowded in and they were not all young. A number were noticed at least twelve years old. The lightest load reported upon the goats was not below 200 pounds. The goats tugged and pulled at the start, and the carriages had to be pushed before they could draw them. "Are you not overworking those goats?" an attendant was asked. "Naw. It ain't hot enough. In July, when it's hot, we give them a rest. But we don't let them stop till their tongues hang out."[14]

The *New York Tribune* thought the answer to "spiritless goats" was not to work the goats less but to work some other animals more. The *Tribune* demanded that the park managers "trot out one of their carefully preserved elephants and give the children a chance to ride it." The only genuine exercise the Central Park elephants ever had, according to the *Tribune*, was "about once a year when the workmen want to move some small building and take one of them out to push it along with his head." The elephants were likely to become "mental wrecks," warned the *Tribune;* a particular elephant named Alice, in fact, was known to "long for the exercise which the carrying of a score of laughing children about the grounds at a leisurely pace would give." The Central Park menagerie did have a serious problem in the confines of its elephant house, but not until a year after the *Tribune*'s call for elephant exercise did that problem burst onto the front pages:

William Snyder, the keeper assigned to Tip the elephant, claimed that Tip had tried to kill him, and the park commissioners thereupon decided that Tip would have to be publicly executed.[15]

Tip was the largest Asiatic elephant in America—nine feet six inches in height, fifteen feet long, and twenty-six feet in girth. He weighed in at five tons, and his age, though variously given as twenty-three, thirty-one, and thirty-three years, was most likely twenty-three. He had been owned at one time by Victor Emmanuel II of Italy, on whose death King Humbert ordered his royal menagerie sold. Tip then passed into the hands of Ludwig of Bavaria, who had been for some time "making inquiries about a young elephant" for his Winter Garden in Munich.

"Professor" Carl Hagenbeck, an animal trainer, bought Tip from Ludwig and later sold him to Adam Forepaugh, whose circus was one of the biggest in postbellum America. Tip performed in Forepaugh's circus from 1883 to 1888; he was the leader of Forepaugh's nine-elephant herd, and some thought him "the best performing elephant in America." On January 1, 1889, when Forepaugh gave Tip to New York City as a gift, he was said to be worth the equivalent of two hundred thousand dollars. Charles Davis, Forepaugh's agent at the time, claimed that the terms of the gift required the city to "keep Tip during his natural life, and neither sell him, give him away, nor kill him," but Forepaugh's actual letter—unearthed by the *Tribune*—said only, "I now take great pleasure in offering you Tip, the second largest of the celebrated Forepaugh herd of elephants, and one of the finest in captivity." To the end, Forepaugh's motive remained in question: had he unloaded on New York City an elephant who had become too difficult for him to handle? Ninety-eight percent of the performing elephants in America were female, but Tip was male, a spectacular "great tusker," with none of the docility that animal handlers counted on in female elephants. Not until well after Tip's arrival in Central Park, however, did reports began to circulate that Tip had killed eight men; the *Herald* repeated these stories but admitted that there was "no good record" of Tip's "evil deeds" in the past.[16]

The wooden elephant house at Central Park was usually described as "frail" and "flimsy." The ceiling was so low that Tip's back brushed it; consequently he was unable to fully raise his head. When William Conklin was superintendent of the menagerie, he had made keeper Snyder take Tip out for a daily walk, but after Conklin retired in 1891, Snyder never exercised Tip again. In March of that year, while Snyder was working in his cage, Tip turned on him and injured him. Claiming that Tip could "break this house into splinters and kill eight or ten persons before he could be captured or killed," Snyder shackled him by one leg. The floor of the cage was not strong enough to hold the shackle, and in February 1892, Tip pulled the bolt out of the floor and charged through the wall of his cage into the next. In 1893, to "protect his keepers," a foot of ivory was cut from each of Tip's tusks; Park Board members took home with them the severed lengths of ivory. By 1894, Tip had not been out of his cage for three years: he was chained to the floor by one forefoot, and "an iron martingale girdled his immense body from which huge iron chains ran to and around his tusks." The shackles were required, according to the zookeepers, "to prevent his raising his head and demolishing things generally."[17]

Tip and William Snyder hated each other. Snyder asserted that each time he entered Tip's cage the elephant tried to kill him with his sole remaining weapon—his trunk. After each such attempt, Snyder had Tip "mercilessly beaten." Knowing that Tip loathed rodents, Snyder took to carrying a cage full of rats into the elephant house and "absentmindedly" leaving it there. He stood by while vicious boys gave Tip apples plugged with cayenne. Snyder could not approach Tip to put more chains on him, as he wished to do; he began feeding Tip by pushing a bucket into the cage and filling it through a hose. Even at that distance, said Snyder, Tip tried to seize him with his trunk. By May of 1894, no human any longer entered the cage in which Tip stood, swaying in his shackles, all day long, every day.

Some thought the "problem" lay not with the elephant but with his keeper and the Park Board. The *Times* questioned "whether so valuable an animal under another keeper might not be made manageable," but noted that if "one or the other had to go," it would certainly be Tip, because "the poor fellow had no pull." William Conklin, former director of the menagerie, said that in his five years' experience with Tip he had always "found him tractable, and saw no reason why he should be killed." Conklin further declared that current menagerie authorities were "not overburdened with knowledge of the animal kingdom." Alvan Southworth, a former elephant hunter, inspected Tip and announced, "If the zoo manager and superintendents would look further than the pig-stys of Hoboken for their expert keepers, such a valuable living example of the mammal species would not be so summarily disposed of by these eminent naturalists—the members of the Park Board of the City of New York." On May 2, 1894, however, the park commissioners took Snyder's side and passed the death sentence on Tip. They deferred the execution for a week to give Tip a chance to attempt "reform." In fact, the commissioners needed a week to discover if anyone knew how to kill an elephant of so great a size.

For ten days, Tip was the top story on every front page, and each development in his case was covered in full detail. At the same time, however, the newspapers unloaded on him the language of irredeemable violence: he was labeled a *villain,* a *rogue,* a *monster,* and a *man-killer,* and he was said to be *wicked, vile-tempered, ungrateful, treacherous, vicious, mutinous, incorrigible, sulky, sullen,* and *doomed.*[18] Tip had no remaining entertainment value: the narrow range of behaviors available to a shackled elephant—swaying back and forth and "twisting fodder into great bundles"—were dismissed as *monotonous* and *tiresome* activities for humans to watch. The *Herald* looked forward to Tip's being replaced with a "sacred albino buffalo from Ceylon, recently imported by Julius Schmitt; its hide is curiously tattooed [an especially appealing feature for a time when human tattoo jobs were rising steeply in popularity] and it is likely to prove a great attraction." The *Herald* found further cause for cheer in the thought that "Tip, as a dead elephant, might be as much of an attraction in the Museum of Natural History as he is alive in the menagerie, while at the same time he could do no harm." The *Tribune,* however, wondered whether the commissioners, "now that they have broken the ice and sentenced a living thing to death," might resolve to "extend their jurisdiction so as to also include all policemen who flaunt pants which bag at the knees." Meanwhile twice the usual crowd visited the elephant house, and Tip himself continued to take food from the hands of little children "with a dignity little consistent with a desire to wreak destruction." Only when he spotted Snyder or one of the park commissioners did he begin to whip his trunk about and twist in his chains.

Between May 2 and May 9, newspaper columns were thick with ideas about, variously, what was "wrong" with Tip, how to save him, how to handle him better, how to improve his living conditions, and how best to kill him. Although it was rumored that keeper Snyder had persuaded the commissioners that Tip needed no exercise, Park Commissioner Clausen nonetheless admitted that Tip's "ill temper" might have been "caused by his being chained in one place for three years without any exercise, but no one had any idea how this monstrous man-slayer could be exercised with safety." When experts attempted to assign other causes to Tip's behavior, available public language about sexuality failed just as completely as it had during the Seeley dinner trial. John W. Smith, director of the menagerie, referred delicately to Tip's "periodical attacks of rebellion," while the *Times* mentioned "a periodical irritation" known as "must"—in other words, sexual excitement—and defined it as synonymous with madness. The *Times* did, however, suggest that Tip's condition might be partially alleviated if the female elephant Juno, who occupied the cage immediately adjacent to Tip's, could be moved to some more distant area "so as to give him a chance to reform." There were hints that "a surgical operation" had been considered in the past, but the thought was abandoned when, as usual, it became clear that no one knew how to perform such an operation.[19] Dr. George F. Shrady, a student of natural history who visited Tip each day, also hid behind a veil when he remarked that Tip was "only laboring under certain natural conditions" and "indicating a desire for fellowship with his mates in an adjoining stall that can deceive no person familiar with the habits of elephants. I think," concluded Shrady murkily, "that the whole trouble has been largely in the direction I describe." In fact, Tip was furiously angry not only because he was unable to raise his head, take a walk, or copulate, but also because he had been beaten for refusing to accept those deprivations. By 1894, Tip was so expressively angry that no one at the menagerie could imagine how to relieve his deprivations without risk to those who did the relieving.

Several persons knowledgeable in animal training suggested that the commissioners seek to hire Eph Thompson, who had been Tip's trainer with the Forepaugh circus and who, "at the end of the act by the herd of which Tip was leader, would seat himself on Tip's tusks and be carried aloft from the ring." But Thompson, said to be "the most successful trainer of elephants ever known," was an African-American working in Europe at that time, and the commissioners never entertained the idea of hiring him. Others suggested that Tip could be electrically managed: Mrs. Donizetti Hamilton Muller of New York City proposed giving Tip "a few acres fenced with bird cage wire, with an electric current around the top. If no man is found courageous enough to take him out, preceded by a band of music and with a few yards of ribbon, I will lead Tip to his new home." Joshua Crosby of 132 West Twelfth Street recommended attaching an electric battery to Tip's hind leg and temporarily paralyzing him with an electric shock whenever he behaved aggressively toward Snyder. Henry Bourne of 307 Madison Avenue, Flushing, pointing out that Tip, at twenty-three, had another seventy years of natural life ahead of him, wondered why, "after having for years deprived him of his liberty and happiness, we cannot do better than to add further to his ill luck in meeting humanity by killing him? If we no longer need him for our pleasure, cannot we present him to himself?" Bourne suggested that a popular subscription of less than two thousand dollars would make it possible to return Tip "to his native wilds and right the wrong of his life." Most curative suggestions, however, were not so benign. Charles Davis, Forepaugh's onetime agent, could suggest only more beating.

> What Snyder ought to do is to get Tip down and hammer him until he squeals. The way to conquer an elephant is to force him to get down on his belly, stretch out his fore legs in front of him and his hind legs behind him, chain them tightly, and then pound him with a hickory sapling about five or six feet long until the elephant squeals. That is their way of acknowledging that they are conquered.

The opinions that carried real weight with the commissioners, however, were those of J. A. Bailey, the circus operator; Hagenbeck, the animal dealer who had sold Tip; and William Newman, former superintendent of elephants with Barnum's circus. All three agreed that Tip ought to be killed at once, "the sooner the better," and the park commissioners unanimously so voted. The controversy then shifted to how the death sentence would be carried out. J. B. Gaylord, an "elephant expert," suggested a method that shrewdly reduced human agency in the death of an animal:

> The victim must be firmly chained, and two of his companion elephants so harnessed that when they are driven in opposite directions they tighten a noose about the condemned animal's neck and choke him to death.

The *Herald* favored "two bullets from heavy rifles," one aimed at the heart and the other at the forehead, a method which it believed to be "merciful and quick." A personage calling himself "Rocky Mountain Dick" wrote to Snyder claiming that he could "kill Tip at a single shot," remarking incidentally that in his time he had "shot enough buffaloes to cover the Park." Snyder himself favored tying Tip to a tree and having a company of militia riddle his body with bullets—if, that is, there was someone who could figure out how to tie Tip to a tree. The *Times* thought that "a wad of gun cotton ignited on his forehead" might work, "but the difficulties with this method would be the attachment of the wad." The commissioners announced that the "general opinion"—among themselves—favored poison:

> Hydrocyanic acid appears to be the surest death agent. One plan is to administer morphine in a large dose in sugared water, and when the stage of narcotism arrives to give a hypodermic injection of hydrocyanic acid behind the left foreleg and as near to the heart as possible. The acid would no doubt act quickly, but no one has been found to tell how the alkaloid [that is, the morphine] would affect a herbivorous animal.

Never before had anyone attempted to kill an elephant with poison. Experiments were called for.[20] The commissioners dropped the morphine idea almost at once, in great part because no one volunteered to deliver the injection; they decided to experiment with the hydrocyanic acid. On May 10, Superintendent William Wallace of the Museum of Natural History visited a downtown drugstore and had eight ounces of hydrocyanic acid made into capsules. That evening, Assistant Menagerie Superintendent Burns and an unnamed physician experimented at the Arsenal with the poison.

> The proceedings were somewhat mysterious, but it is known that two guinea pigs, a vial or two, and some cotton wool made up the "property." The pigs were dead when it was all over, but their lamentations were loud while the performance was in progress. In fact, one of the victims, which was thought to be dead, displayed a surprising amount of life when the doctor questioned it with a lancet. Sergeant Mulholland, who was at the desk in the police station downstairs, had this to say about it: "If them pigs made all that fuss after eating some prussic acid in a bit of carrot, what will that five-ton elephant do when they tackle him in the morning?"

About the guinea pigs the commissioners said nothing in public; they pronounced hydrocyanic acid unsuitable "because it would create a danger to those holding the autopsy on Tip and preparing his skeleton and skin for the Museum of Natural History." Their chosen poison, they announced, would be the "safer" potassium cyanide—a poisonous compound which is, in its effects, indistinguishable from hydrocyanic acid and exactly as dangerous on contact.[21] The commissioners further announced that the actual time of Tip's execution would not be made public, "as a crowd is one of the things that is not wanted." The latter pronouncement denied the public's interest in knowing about, if not seeing, the event; the commissioners could not get away with it any more than could Mrs. Bradley Martin when she tried to prevent street crowds from seeing her costumed ball guests, or Herbert Barnum Seeley when he tried to conceal Little Egypt in an upstairs room at Sherry's. An outpouring of public scorn forced the commissioners to reveal that on May 11, at exactly six o'clock in the morning, keeper Snyder would feed three carrots to Tip, thus whetting his appetite and "tending to quiet any suspicion which the early gathering [of invited spectators] might arouse." In carrot number four would be concealed capsules containing two ounces of prussic acid. All unawares, Tip would crunch down on the fourth carrot. At that point, "if the experts are correct, Tip will know no pain. A giant hand will clutch his heart. His respiration will cease. His limbs will shoot out convulsively and stiffen in an instant, and he will totter and fall to the floor—lifeless." The entire event would, the experts predicted, be over in no more than three seconds. Assistant Superintendent Burns issued invitations to the death show:

> Department of Public Parks
> Office of Menagerie, May 10.
> Admit bearer to execution of Tip. G. R. Burns.

Meanwhile Superintendent William Wallace of the Museum of Natural History, a man "whose feelings concerning the execution were but a short remove from joy," dispatched a photographer to the elephant house to take snapshots of Tip in "lifelike poses." Wallace planned, in fact, to fully separate Tip's skin from his skeleton and thus display two Tips in the museum—a skeletal Tip and a stuffed Tip. The photographs would be consulted "when the Museum was ready to begin the work of stuffing and mounting" Tip. The photo session was not successful: the low beams of Tip's cage interfered with the photographer's getting a full view, and Tip refused to stop swaying and stand still for his photo.

On May 10, one day before the planned execution, Tip's keeper "expected that a crowd would annoy him," so Assistant Superintendent Burns closed the elephant house, had the doors barricaded, stretched ropes across the approaches, and ordered park policemen to "keep back the crowd." At noon, "the hill behind Tip's house was thronged by more than five hundred people, and nearly a thousand more filled the walks about the cages of the lions and tigers." Some onlookers critiqued the "cowardly" commissioners' execution plan, others waited in hopes of viewing the execution, and some wanted to see Tip knock down the elephant house and "run amuck." Tip himself was heard to trumpet occasionally; only when Park Commissioner Nathan Straus and a party of his friends visited the elephant house in company with a few keepers did "their presence seem to enrage him instantly. He trumpeted loudly and threw the weight of his gigantic shoulders against the iron bars which make his prison. The heavy framework creaked." Straus and his friends left at once.

The unhappy and physically miserable Tip had been subjected to some of the late nineteenth century's most painful human social ideas, none of which adequately described or answered to his situation. He had publicly expressed sexual need in a society that denied such expression even while it silently acknowledged sexuality in certain male persons. Because Tip had no power but the physical and no money, he was seen as having no special right to space: his cage was, according to the *Herald,* about the size of a hall bedroom in a tenement. In a society that had had significant experience with panics, Tip's human keepers made him an object of fear when they claimed that at any moment he might break loose of his restraints, behave unpredictably, and bring about panic. He had also been subjected to the notion that protesting intolerable conditions constituted a type of incurable madness. When human beings went "mad," they were incarcerated and whisked out of public view, but Tip was already incarcerated and no one knew how to whisk him away. And finally, because his "temperament" was not under the kind of control that humans demanded of each other, the authorities— conveniently assuming reform to be possible for anyone—gave Tip a week to "reform." No one, however, considered how to render that ultimatum intelligible to him. Human problems with Tip were insoluble in human terms, and so humans decided that Tip would have to be removed from the terrible distress of his life.

The day of execution, May 11, exposed the deep connections between dominance and fear—men's need to gain dominance over Tip, their fear of failure, and their further fear of being harmed in their drive for success. Before five in the morning, a crowd began to assemble outside the elephant house—park officials, newspapermen, doctors, keepers, policemen, and interested citizens. Snyder, looking "fretful and anxious," was there, as were several persons who held Snyder responsible for the commissioners' decision to execute. Tip himself, realizing something was afoot, could be "seen through the open window of his dungeon, swaying to and fro with unusual vigor." Then a squad of a dozen policemen, two abreast, marched to the elephant house and unlocked the doors. Only white males were allowed to enter the "death chamber"—first the doctors and menagerie officials, then the newspapermen, each presenting his pass to the police censor.

Shortly before seven o'clock, the enthusiastic William Wallace of the Museum of Natural History arrived, carrying a magazine rifle, which inspired among the newspapermen considerable sarcastic comment as to the "dangers of hunting elephants in the Indian jungles." At a table off to the side, Burns and one of the doctors were busy "boring holes in the apples and carrots which were to contain the deadly capsules." Each vegetable was plugged with two ounces of cyanide, and each, it was announced, "should cause death in a very few moments." Meanwhile the densely packed crowd of about a hundred men pressed close to Tip's cage and "a great trunk was thrust quietly out between the bars until its prehensile tip almost tickled the man who stood nearest."

> There was a momentary stampede, and Dr. George Huntington of the Vanderbilt clinic announced, "Gentlemen, you will remain here entirely at your own risk." Everyone seemed willing to do that, and a few advised the doctor to go ahead with the show. "This elephant," he continued, "is quite likely to spit out a highly dangerous substance. If one drop of it alights upon a mucous membrane, say that of the eye or lips, it will mean certain death." Three park policemen, who were not on duty, promptly left the building. The majority, however, pulled their hats well down over their eyes, turned up their collars and replied, "All right; we'll risk it."

Evidently that response to Huntington's announcement was entirely unexpected. Huntington's little speech had been supposed to frighten the newspapermen out of the building; the invitations and press passes had been a sham. When their scare tactics failed, park commissioners and menagerie officials, who were determined to have the show to themselves, ordered the park police to clear the room, the steps in front of the building, and the adjoining fences. Newspapermen ejected from the elephant house fought for vantage points on the grass outside the open back of Tip's cage.

Assistant Superintendent Burns then handed the first of the plugged carrots to keeper Snyder, but he refused to deliver the carrot to Tip. Burns then turned to Otto Mopis, a "careless-looking, red-faced man" said to have worked for Barnum and to be acquainted with Tip, even though Tip himself gave no sign of recognizing him. Mopis stepped up to the bars and held out the carrot. "Here you are, Tip," he said. In preparation for this moment, Tip had been deprived of food for the previous twenty-four hours; Burns rendered this information by announcing that Tip "had been fasting."

> Tip accepted the carrot readily enough, felt it thoughtfully with his trunk and stowed it away in his mouth. He spat it out a moment later with a snort of disgust, which sounded like steam from an escape valve. The carrot was broken. A part of the poison, Dr. Allen said, had been retained.

The silence of the spectators was complete and intense: "a shudder went through the onlookers as they expected to see, in a moment or two, the enormous mass of flesh crash in a helpless mass to the floor." Instead, Tip began to spit out pieces of the carrot, along with quantities of saliva and a bright white foamy fluid. Tip stamped on the mess on the floor. At Burns's order, four men armed with Winchester rifles filed into the room "and took their stand near the cage, ready to shoot Tip if he should break loose." Mopis then offered Tip a poisoned apple. Tip sniffed it, rubbed it against the underside of his trunk, threw it down, and stamped on it, whereupon the plug, which had been fastened with a pin, fell out.

At this point, everyone's attention was distracted when a well-dressed man said to be Dr. Wright of West Eighty-first Street, an invited guest, entered the elephant house armed with a rifle. A few moments later he reappeared outside, *sans* rifle. Word passed through the crowd that Wright was "suffering from a nervous attack" and had been disarmed and ejected when he "acted so strangely as to attract attention" and pressed the muzzle of his weapon against the back of a spectator.

Meanwhile Tip crushed and swallowed a poison-free loaf of bread intended as "reassurance," but the next poisoned carrot Mopis offered him he "tossed away contemptuously." Then he became horribly ill, no doubt from the contents of the first broken carrot. He backed up into the angle formed by the north wall and the bars of his cage and, his legs shaking, leaned heavily against the wall. His trunk hung straight down, motionless. Two more apples and two loaves of bread he crunched under his feet.

> At 7:32 o'clock, Tip began to retch violently, but his attempts to vomit were ineffectual, and the sounds he made in endeavoring to rid himself of the poison in his system were painful to hear. The poor brute seemed to be in great pain.

Four times "his great body gathered in spasms and he bent forward with something like a gasp." Commissioner Nathan Straus went outside and said hopefully to the crowd, "He's a little sick." After each spasm, however, Tip rallied; by eight-thirty he had begun to swing his trunk again, and by nine o'clock "he was gathering up bundles of grass and pelting the rear wall of the pen with them." Burns offered Tip an "unloaded" potato and a carrot, but Tip refused to eat. The disappointed Dr. Horace Allen, curator at the Museum of Natural History, insisted that Tip must have ingested, with the first broken carrot, enough poison "to kill several hundred men almost instantaneously." William Wallace, tired of waiting for his prize, said it was time "to chalk Tip's forehead and send a volley of bullets at him," but President Hankinson of the SPCA would not permit shooting. Commissioner Straus, looking disgusted, announced that the "execution would be postponed indefinitely," the commissioners departed, and the park police went to work dispersing a crowd that had grown into the thousands.

Wallace, Burns, and Snyder were not, however, about to be stopped; disregarding the commissioners' directive, they reentered the elephant house on their own and began to plug with cyanide every elephant food they could lay their hands on. They were, as the news-papermen saw it, determined to make Tip commit suicide. Burns poi-soned more carrots and apples, all of which Tip refused and trampled on, and then went to work on peanuts. Burns removed the nut from each shell, inserted a cyanide capsule in its place, and then "mended the shell so neatly that it seemed well nigh perfect." At noon, keeper Snyder, free of whatever compunctions had held him back from hand-ing Tip the first poisoned carrot of the morning, held out a handful of poisoned peanuts to Tip.

> The elephant approached them very gingerly. He hooked up two or three and held them for a full minute, waving his trunk in great circles. Then he dropped them, capsules and all, and blew out his breath as if to be sure that none of the shells was retained.

For three hours more, Burns and Snyder kept trying and failing. Then Snyder made a paste from a pailful of bran mixed with water, and he and Burns embedded in it two large gelatine capsules and thirteen small ones, about three ounces of cyanide altogether. At seven minutes past four, they delivered the bran to Tip, and he ate it. Four minutes later, he "gave a mighty gasp for breath," his back arched, his legs quiv-ered, and he went into convulsions. He trumpeted in fear, swung about, twitched convulsively, and "screamed in terror and pain."

> As if in a desperate effort to inflate the mighty lungs which the fumes of the poison had paralyzed, he reared and threw his forehead against the solid timbers of the wall. Then the dying elephant wheeled in his tracks and charged for the rear door of the pen. His body went on, but the big chain which made him doubly a prisoner drew one foreleg back beneath him and he went to his knees. He knelt there for a moment— only a moment—and then fell over on his side, reddened by a hemorrhage from the lungs.

Snyder entered the cage first and pronounced Tip to be dead; he and Burns stood over Tip to have their picture taken, while Wallace busied himself assuring everyone that Tip, angry to the last, "must have bitten his tongue in his rage, and that accounted for the blood." Within half an hour, John Rowley, chief taxidermist of the Museum of Natural History, arrived with a force of ten men, to whom Wallace distributed long knives, "and well before midnight they made great progress in flaying the huge carcass."[22]

Reports of Tip's death agony were juxtaposed with detailed descriptions of the value of his corpse to "science." The *Tribune* found further refuge in standard American ideology when it suggested that "whether Tip's death was absolutely necessary for the preservation of human life will always be a matter of individual opinion." The *Herald* bitterly pronounced the dead Tip to have become, finally, "a good elephant," but also gave credence to Commissioner Bell's postmortem speech to the newspapermen. Just as it had been important at six in the morning to represent Tip as irredeemably violent, so it was important at four-thirty in the afternoon to insist that the elephant had not suffered, that it was men who had been threatened and men who

The Dead Rogue.

had suffered. In service of this notion, Bell, standing beside Tip's body, seized the occasion "to correct an impression that he feared might become general—that Tip had been tortured from early morning up to the time of his death." Nothing of the kind had occurred, according to Bell; Tip had been "ill" for just twelve minutes. The *Tribune* and the *Herald* played along, but the *Times* was having none of it, as is clear from its May 12 edition:

BIG ELEPHANT TIP DEAD
Killed with Poison After Long Hours of Suffering.
Terrible in His Death Agony, He Burst His Chains

WAS ON THE POINT OF BREAKING FROM THE BUILDING WHEN HE SUCCUMBED
TOOK CYANIDE OF POTASSIUM IN WET BRAN
HAD REJECTED POISONED APPLES AND CARROTS
SNYDER GAVE HIM LAST CAPSULE

Tip is dead. The big elephant that had been for years the playmate of the children in Central Park, and the terror of Snyder, who was his keeper, was put to an agonizing death yesterday. The largest captive of his kind since Jumbo paid with his life the penalty of refusing to tame his elephantine nature at the command of a man he hated. Tip's last day of life was full of misery, of pain, of torture. For more than nine hours the poison was corroding his vitals. It is claimed for this poisoning that it was a success. Its success lay in the fact that Tip was killed. He took his first poison at 6:58 o'clock in the morning. It was 4:19 o'clock in the afternoon when he fell dead. What he suffered during those intervening hours is a matter of conjecture.

It was Tip's misfortune that in his prime he fell into the hands of powerful and demanding human males. Like others of their time, Tip's controllers demanded human satisfaction from the nonhuman, they required public silence on the subject of sexuality, and they were, ominously for him, fond of taxidermy. It was Tip's further misfortune to have arrived on the cityscape during the greatest collectors' moment in American history. Framed in that moment, Tip was but one item in a large collection and as such was subjected to standard collectors' behaviors: any collector, after all, felt occasionally impelled to edit the collection, and suddenly lost interest in this or that object, and engaged regularly in replacing and upgrading elements in the collection. Furthermore, Tip had got too big for the collection. He had dangerously outgrown his allotted display space and had had to be fastened to his shelf. Instead of catering to the needs of the collector, he had come to have needs of his own. Most unsettling of all, he had apparently come to loathe his current collector, and to communicate that loathing to every visitor who dropped by to view the collection. Within the collecting framework, he looked ripe for deaccessioning.

City dwellers in the 1890s employed a variety of methods for managing desirable but hard-to-control objects. None of their usual methods, however, seemed appropriate to Tip. He could not be distributed as leftovers to the poor, the method used by the rich to dispose of excess goods and palliate their critics. He could not be sold at auction, a disposal method that such collectors as William Merritt Chase resorted to when their acquisitiveness got the better of them and they felt overwhelmed by their artgoods. The tactic of removing troublesome items from view was the Bradley Martins' favorite: they locked up jeweled wedding gifts too dangerously valuable to be put on display, and they attempted to control public reaction to their entertainment excesses by elaborately closing them off from public view. Publicly troublesome humans found on city streets were regularly jailed or and tucked out of sight on Blackwells Island. Similarly, Herbert Barnum Seeley, hot to titillate his party guests with the sight of an exotic woman in shackles, tried, when scandal broke, to tuck himself out of sight. Then he tried unsuccessfully to pay off the woman—Little Egypt—and tuck her out of sight. Eventually it required the efforts of a police headquarters hearing room full of men to whisk her out of public view.

Tip could not be paid off and he could not be silenced; unlike Little Egypt, he could not be sent off to trumpet in private to a grand jury. He was already locked up, he was already shackled, and no one knew how to remove him from public view. Although Tip had been offered a sham option of "reforming," he in fact had no options whatever, and his controllers persuaded themselves that they had but one option; that one, however, they would have to exercise in public. Tip had been immobilized and incarcerated in public, and his controllers could not conceal from the public what they were prepared to do to him in order to transform him into a "good elephant." They were going to disassemble him, just as they might choose to disassemble an out-of-style cozy corner by knocking it down into its component parts of fabric, plywood, and stuffing. It was a process they had already begun three years earlier when they sawed off his tusks and distributed them to collectors on the Park Board, and they finished the process when they dismantled Tip: his skin they treated as a fabric, his skeleton they disarticulated, and his toxic innards they threw away. Disassembled, one bad Tip could leave the Central Park menagerie collection and become—or so they hoped—two good Tips in the Museum of Natural History's collection. Dead, Tip could decorate a collection as quietly as a stuffed hummingbird decorated a chiffon ruche; he could be as fascinatingly silent as the Peruvian mummy head that stuck in the memory of everyone who saw it at William Merritt Chase's studio.

It is unlikely that Tip's controllers, when they laid their plans to deaccession him, were as ignorant as some of their critics thought them to be at the time. They knew Tip was going to suffer publicly and at length; their experiments on guinea pigs had told them so. Their final effort to deny that he had suffered, and to deny it in front of a crowd that had watched him retch his lungs out, was so unconvincing as to raise questions about powerful sensation-seekers and the nature of their desires. Tip's suffering had been, possibly, no mere series of ghastly blunders but an actual source of excitement and fascination. As such, it required from them the silence they were accustomed to draping over all of their sensational excitements.

American city dwellers in the 1890s saw a great deal of suffering. They needed to develop ways to deal with it, and they did develop those ways. The late-twentieth-century soap-opera viewers' cliché that one watches the sufferings of others in order to feel better about one's own miserable life is too facile for the complications of a social situation in which Americans watched not actors miming suffering but real living sufferers. Those who went out of their way to view suffering revealed most vividly the nature of the sensations they desired. They cultivated insensitivity to the physical and mental discomfort of other beings, and then found thrills in viewing that discomfort. They made, for example, a nervous jaunt and a "social study" out of the voyeuristic thrill of staring at sleeping down-and-out men in flophouses. To experience the special thrill of withholding pleasure, they lingered in toy departments and stared at poor urchins lined up, hands clasped behind their backs, studying the expensive toys that luckier children would get from Santa. Some of them liked to tease: Mrs. Bradley Martin loved to tease adults with detailed reports of lavish entertainments that they were not to see and luxury foods that they were not to eat. Philanthropists of all sorts teased malnourished children by separating them from their hungry parents, treating them to one big meal a year, extracting from them orchestrated expressions of gratitude, and then dispatching them back into deprivation.

Tip had been officially pronounced to be an "ingrate": in the parlance of the time, that meant he was a grown-up who proclaimed his suffering, who defined it as caused by forces outside himself, and who insisted that it need not be. At five tons, Tip was also the largest and most visible of those who suffered through the 1890s. Tip was too big to be pacified with champagne and too mature to be distracted by toys. He became angry when he was teased, remembered every man who had teased him, and waited for his chance to repay those men. Unlike the children who sent up cheers for the gift of a single dinner in a long year, and unlike the sleeping down-and-outers who never knew that slummers were peeking at them, Tip was a vocal social critic. He indicated his opinion of the powerful in his life, and as such he was dangerous: he took his place with George Roeth shooting up the ceiling of Delmonico's, and Camille Rheinhardt pounding on John D. Rockefeller's front door, and Little Egypt in the witness chair launching vivid narratives about socially important drunken unbuttoned men. The men who controlled Tip could not return him to childhood, the state favored by the powerful who wished to demonstrate easy control over others, but they could make him as needy as the submissive children they treated to yearly dinners, and they did. Then they proffered him their specially concocted once-in-a-lifetime meal, and because he was very hungry, he ate.

NOTES

1. In 1886, the New York state legislature passed antisparrow legislation: "The English or house sparrow (*Passer domesticus*) is not included among the birds protected by this act and it shall be considered a misdemeanor to intentionally give food or shelter to the same." In 1893, the SPCA countered by distributing in primary schools a "catechism on kindness to animals," from which comes the following (*New York Herald*, 4 December 1893):

Q. Is the sparrow a native of this country?
A. It is so now, but only a few years ago there were no sparrows in America.
Q. Why were the sparrows brought to this country?
A. Because the insects were killing so many trees that the sparrows were needed to destroy the insects.
Q. Did the sparrows save the trees?
A. Yes; since the sparrows were brought over the trees have been saved.
Q. In winter time, when there are no insects and the snow is on the ground, does not the sparrow have a hard time?
A. Yes, he has a very hard time, and many of them die of hunger.
Q. What can we do for the sparrows in winter?
A. We can give them bits of bread.

2. *New York Herald*, 7 February 1897, 2 December 1894, 8 December 1895. In 1899, women's hats were decorated with the quills of the bald eagle, and "a great hue and cry went up against a possible slaughter of the national bird for the purpose of woman's head decoration" (*New York Herald*, 3 December 1899). In *Men, Beasts, and Gods*, Gerald Carson writes: "The millinery trade fought off critics with arguments that have a familiar ring. Wild bird feathers supported a legitimate business. They provided the livelihood of hard-working Americans. And the birds did a lot of damage. . . . Those who raised questions were clearly cranks, busybodies, and probably socialists. If the woman who was tempted by the nuptial plume of the egret voiced a scruple, her milliner had a story ready to satisfy her qualms. The plumes had been molted. They were obtained from game birds only. And so on" (184).

3. *New York Herald,* 26 July 1893.

4. *New York Herald,* 2 March 1893.

5. *New York Herald,* 16 February 1895, 21 January 1894, 28 January 1894, 27 December 1896.

6. *New York Herald,* 1 December 1895, 8 December 1895. In *Men, Beasts, and Gods,* Gerald Carson notes: "Though the need for furs is today as obsolete as the human tail, rare skins remain a symbol of conspicuous leisure and expensive adornment. A considerable number of Venuses in furs experience sexual excitement from contact with the skins of beasts, a phenomenon upon which Dr. von Krafft-Ebing makes some interesting observations under the title *Zoophilia Erotica* in his classical study of perversions" (183).

7. *New York Herald,* 20 January 1896, 23 December 1894.

8. *New York Herald,* 25 February 1894; Hardman, 34–35.

9. *New York Herald,* 4 December 1893; *New York Tribune,* 3 May 1894.

10. McAllister, 353–354. All material on the Midnight Band of Mercy is drawn from the *New York Herald,* 13 September 1893–18 November 1893.

11. Henry Bergh was the founder of the American Society for the Prevention of Cruelty to Animals. While there was no place in the SPCA for women, the Midnight Band of Mercy's membership was all women and their activities were opposed by the SPCA. In *City of Eros,* Timothy Gilfoyle discusses the preventive societies whose membership included primarily prominent Protestant and Republican men. Each society "concentrated on a specific social vice, envisioned itself as an enforcer of the law, and believed that the spread of social deviancy justified extralegal action." Preventive societies "adopted their own private, unregulated methods" and "won vague law-enforcement responsibilities from the state, thereby acquiring a quasi-governmental power other reform groups never enjoyed. . . . The absence of female participants also distinguished preventive societies from their reform contemporaries" (186–190).

12. *New York Tribune,* 3 May 1894. Gerald Carson in *Men, Beasts, and Gods* notes that "the idea of licensing cats comes up from time to time, but never gets anywhere. It is doubtful if anyone truly owns a cat. And besides, there just isn't any place on a cat where a license tag can be securely attached" (145).

13. Rosenzweig, 344; Rae, 48; Strange, 44.

14. *New York Herald,* 8 May 1893, 10 May 1893; *New York Times,* 6 June 1893.

15. *New York Tribune,* 19 April 1893. All material on the elephant Tip is drawn from the *New York Herald,* the *New York Times,* and the *New York Tribune,* 5 May 1894–14 May 1894.

16. Blunt, 92. Forepaugh's circus toured Europe in the mid- to late 1870s, during which period Forepaugh may have seen or heard of Tip. Young elephants were much sought after for royal menageries in Europe, undoubtedly because they presented none of the "problems" associated with sexually mature elephants. When baby elephants were no longer babies, their royal owners disposed of them to circuses.

17. From a certain point of view, it might be said that Tip spent the last three years of his life offering to men the covert sexual entertainment they apparently experienced when looking at a shackled being—the same entertainment Herbert Barnum Seeley sought when he asked Little Egypt to appear shackled at his dinner in 1896. As in the cases of Salon art's barefoot peasants and of *The Thousand and One Nights*, the sexual content of the text, so to speak, made itself available to those who were looking for it.

18. A monster is a being that has no comfortable social place; in his short story "The Monster" (1898), Stephen Crane creates the horribly burned Henry Johnson as just this sort of excluded—and damaged—figure. In *Rubbish Theory*, art historian Michael Thompson discusses "monster exclusion" as a "distinct, and often dominant intellectual style" that is, "at its worst, intolerant, puritanical and repressive. At its best, it reveals a dubious prettifying intent that leads to the pretense that things are tidier than they really are" (132).

19. The code for medical and surgical matters involving sex and reproduction was the absence of any code: thus an abortion was always referred to as "an operation," and no other procedure was so called. The "surgical operation" contemplated for Tip must have been castration.

20. The late nineteenth century combined a ready availability of drugs and poisons with both extreme carelessness and a reckless take-a-chance attitude. If an individual wishing to do in another individual sent to him or her, through the mails, a box containing unlabeled vials and powders, there was at least a fifty-fifty chance that the recipient would ingest the goods. Moreover, household poisons and prescription medicines were regularly mixed up, with catastrophic results. On 29 August 1893, for example, Alderman J. A. Arnold of Newark, New Jersey, after "suffering severely for some days with two carbuncles on his neck," had the carbuncles lanced by a physician, who "left him a small bottle of tonic." At about 7:30 P.M., Arnold told his wife to pour out a teaspoonful of the tonic. "The bottle stood on a shelf in the kitchen beside another, and although it was quite dark in the room Mrs. Arnold thought she took down the right one. Her husband swallowed the dose she gave him, but immediately exclaimed, 'My God! You have given me the wrong stuff. It's car-

bolic acid.' " Fifteen minutes later he was dead. Stories like his were so ordinary that the newspapers became bored with them. On 10 March 1895,
under the headline TOOK THE WRONG MEDICINE, the *Herald* explained "why
this headline so often appears in the daily newspaper. It is an odd trait in
human nature that a man who has been ordered by his physician to take pare-
goric will never take it if there is any carbolic acid or prussic acid in the house
that he can absorb in preference. Statisticians who have studied the thing
declare that an invalid will search the whole house for a poisonous drug and
drink it rather than the medicine ordered by the doctor. It isn't only children
who make these blunders. Doctors will tell you that they have only to label a
bottle 'Lotion: for External Applications Only' to make sure of its being
drunk. If a patient gets a bottle of corrosive sublimate to put on a felon on his
great toe and doesn't use it all he will carefully save it. Ten years afterward a
doctor gives some cough mixture to him, and then he goes and hunts up the
corrosive sublimate bottle, plays three card monte with it and the cough mixture, gets them thoroughly mixed up so that he can't tell one from the other,
and then when he feels that tightness across the chest that the doctor told him
about he swallows a part of the corrosive sublimate and leaves his widow to
collect the life insurance. By no accident is the cough mixture ever taken—it is
always the corrosive sublimate."

21. Potassium cyanide, according to the current Laboratory Chemical
Safety Summary on the subject, is a severe poison and can be fatal if swallowed, inhaled, or absorbed through the skin: "Inhalation: Corrosive to the
respiratory tract. The substance inhibits cellular respiration and may cause
blood, central nervous system, and thyroid changes. May cause headache,
weakness, dizziness, labored breathing, nausea and vomiting, which can be
followed by weak and irregular heartbeat, unconsciousness, convulsions,
coma and death. Ingestion: Highly Toxic! Corrosive to the gastro-intestinal
tract with burning in the mouth and esophagus, and abdominal pain. Larger
doses may produce sudden loss of consciousness and prompt death from respiratory arrest. Smaller but still lethal doses may prolong the illness for one or
more hours."

22. In 1894, the American Museum of Natural History had on display the
remains of the elephant Sampson, who burned to death at Bridgeport in 1887,
and those of the elephant Jumbo, who was killed in a railroad accident at St.
Thomas, Ontario, in 1885. William Wallace saw Tip as a valuable addition to
this elephant display: he expected eventually to exhibit both Tip's articulated
skeleton and a stuffed version of Tip, "fully mounted with replacement tusks
and with glass eyes, so it will look as the elephant did in life." It is impossible

to know in how many ways the projected procedures went awry. Tip's hide was at some points an inch and a half thick, and so "peeling him," as the *Tribune* called it, was difficult. Moreover, when he died, he dropped onto his left side and how he might be turned over, within the cramped confines of the cage, was unclear. The "experts" planned to cure the skin in "a vat of preservative fluid for several months" and then stretch it on frames; the bones would be scraped and bleached. The skeleton of Tip (*Elephas maximus*) is today item #3819 in the collection of the American Museum of Natural History; accession records indicate that it was given to them on 12 May 1894, by the Parks Commission. It has never been displayed.

In 1927, Tip's skin passed into the possession of the Yonkers Museum, now the Hudson River Museum. There, in 1929, the stuffed Tip went on display. He stood in a narrow room (originally the dining room of the Trevor Mansion, where the museum was lodged). The tusks on his mount were not the originals; although in 1894 the Museum of Natural History had hoped that Park Board members who had taken home the tusks sawed from the living Tip might return them for attachment to the proposed stuffed Tip, apparently the board members did not comply. By 1956, Tip had become an obstacle to the functioning of the Yonkers Museum's new Spitz projector that would cast views of the galaxy onto a muslin dome installed on the ceiling of the dining room. Museum trustees did not know what to do with Tip; the curator suggested cutting off Tip's head, putting it on display, and disposing of his body.

Meanwhile, in Somers, New York, local residents on the Somers Circus Museum Committee heard about Tip and applied for possession; they envisioned him as the centerpiece of their projected Circus Museum. The museum trustees in Yonkers thought the giveaway to be a fine idea. No one, however, could remember how Tip had got into the museum, and in order to get him out, doorways and floors had to be dismantled. Finally, the Murphy brothers, movers from Mt. Kisco, were able to extricate Tip and load him onto a flatbed trailer truck. On this he rode backward but standing to Somers, a distance of forty miles. The necessity of dodging tree branches, traffic lights, and low bridges made it a slow and difficult journey. When the truck arrived at Somers, eight hundred school children released from class cheered for Tip, and the Somers Central School band played several tunes. Somers had three municipal buildings—a town hall (which locals were already calling the Elephant Hotel), a highway garage, and a firehouse. Just as at the Central Park zoo, however, Tip was too big for the quarters that humans had selected for him; he would fit into none of the three buildings and so was transported four miles out of Somers to Granite Springs. There, lodged in a barn owned by

Otto Koegel, a New York lawyer and the Somers town historian, Tip was to await the building of the Somers Circus Museum. The town of Somers began to sell shares of stock in Tip, at a dollar apiece, in order to raise money to build the museum, but apparently the effort never came to fruition. Meanwhile, because Koegel's barn was a shambles of a building with a bad roof, the weather began to leak in on Tip and he deteriorated. In 1962, he was taken out of the barn and put on a railroad siding where, according to local observers, he "just sank into the ground." The Somers Historical Society retains the pair of gold-banded tusks that were attached to Tip when he moved to Somers.

For their aid in tracking the remains of Tip, my thanks go to Patricia A. Brunauer, Department of Mammalogy, Division of Vertebrate Zoology, American Museum of Natural History; to Laura Vookles, Curator of Collections at the Hudson River Museum; to Cecelia Becker, Reference Librarian at the Somers, Connecticut, Public Library; and to Fred Dahlinger, Jr., Director, Collections and Research, Circus World Museum, Baraboo, Wisconsin.

BIBLIOGRAPHY

Chicago Tribune
New York Herald
New York Journal
New York Sun
New York Telegram
New York Times
New York Tribune
St. Louis Post-Dispatch

Adams, W. E. *Our American Cousins*. 1883; rpt. Lewiston: Edwin Mellen Press, 1992.

Afshar, Haleh, ed. *Women in the Middle East*. New York: St. Martin's Press, 1993.

Archer, William. *America Today*. New York: Charles Scribner's Sons, 1899.

Armstrong, David Maitland. *Day Before Yesterday*. New York: Charles Scribner's Sons, 1920.

"Artist-Life in New York." *Art Journal,* 1880, 57–58, 121–123.

"Art-Notes from Paris." *Art Journal,* 1880, 379.

Auchincloss, Louis. *The Vanderbilt Era: Profiles of a Gilded Age*. New York: Scribner's, 1989.

Aveling, Edward A. *An American Journey*. New York: John W. Lovell, 1887.

Badeau, Adam. *Grant in Peace*. 1887; rpt. Freeport, N.Y.: Books for Libraries Press, 1971.

Baker, Paul R. *Stanny: The Gilded Life of Stanford White*. New York: Free Press, 1989.

Balsan, Consuelo Vanderbilt. *The Glitter and the Gold*. 1952; rpt. Maidstone, Kent, England: George Mann Books, 1973.

Banta, Martha. *Imaging American Women: Idea and Ideals in Cultural History 1880–1920*. New York: Columbia University Press, 1987.

Barron, Clarence W. *They Told Barron: Notes of the Late Clarence W. Barron*. New York: Harper & Brothers, 1930.

———. *More They Told Barron*. New York: Harper & Brothers, 1931.

Bates, E. Catherine. *A Year in the Great Republic*. 2 vols. London: Ward & Downey, 1887.

Beadle, Charles. *A Trip to the United States in 1887*. Privately printed, 1887.

Bennett, Arnold. *Your United States: Impressions of a First Visit*. New York: Harper & Brothers, 1912.

Berry, C. B. *The Other Side: How It Struck Us*. New York: E. P. Dutton, 1880.

Bigelow, Poultney. *Seventy Summers*. New York: Longmans, Green, 1925.

Birmingham, George A. [pseud. James O. Hannay]. *From Dublin to Chicago: Some Notes on a Tour in America*. New York: George H. Doran, 1914.

Bisland, Elizabeth. "The Studios of New York." *Cosmopolitan* 7:1 (May 1889), 3–22.

Blaugrund, Annette. *The Tenth Street Studio Building*. University of Washington Press, 1997.

Blouet, Paul [pseud. Max O'Rell]. *Jonathan and His Continent: Rambles Through American Society*. Trans. Mme. Paul Blouet. New York: Cassell, 1889.

———. *A Frenchman in America*. New York: Cassell, 1891.

Bose, Sudhindra. *Fifteen Years in America*. Calcutta, India: Kar, Majumder, 1920.

Bourget, Paul. *Outre-Mer: Impressions of America*. New York: Charles Scribner's Sons, 1895.

Brandon, Ruth. *The Dollar Princesses: Sagas of Upward Nobility 1870–1914*. New York: Knopf, 1980.

Bridge, James H. [pseud. Harold Brydges]. *Uncle Sam at Home*. 1888; rpt. New York: Arno Press, 1974.

Brooks, John Graham. *As Others See Us: A Study of Progress in the United States*. New York: Macmillan, 1908.

Brown, Elijah [pseud. Alan Raleigh]. *The Real America*. 1913; rpt. New York: Arno Press, 1974.

Browne, Junius Henry. "The Bread and Butter Question." *Harper's Monthly* 88 (December 1893), 273–280.

Buckley, Michael B. *Diary of a Tour in America*. Dublin, Ireland: Sealy, Bryers & Walker, 1889.

Burne-Jones, Philip. *Dollars and Democracy*. New York: D. Appleton, 1904.

Burnley, James. *Two Sides of the Atlantic*. London: Simpkin, Marshall, 1880.

Burns, Sarah. *Inventing the Modern Artist: Art and Culture in Gilded Age America*. New Haven: Yale University Press, 1996.

Burrows, Edwin G., and Mike Wallace. *Gotham: A History of New York City to 1898*. New York: Oxford University Press, 1999.

Carlton, Donna. *Looking for Little Egypt*. Bloomington, Ind.: IDD Books, 1994.

Carson, Gerald. *Men, Beasts, and Gods: A History of Cruelty and Kindness to Animals*. New York: Charles Scribner's Sons, 1972.

Castellane, Boni de. *How I Discovered America*. New York: Knopf, 1924.

Churchill, Allen. *Park Row*. New York: Rinehart, 1958.

Cikovsky, Nicolai. "William Merritt Chase's Tenth Street Studio." *Archives of American Art Journal* 16 (1976), 2–14.

Clark, Kenneth. *The Nude: A Study in Ideal Form*. New York: Doubleday, 1956.

Clews, Henry. *Fifty Years in Wall Street*. New York: Irving Publishing, 1908.

Collins, Frederick L. *Glamorous Sinners*. New York: Roy Long & Richard R. Smith, 1932.

Committee of Fifteen. *The Social Evil, with Special Reference to Conditions Existing in the City of New York: A Report Prepared in 1902 Under the Direction of the Committee of Fifteen*. 2nd ed., rev., with new material, ed. Edwin R. A. Seligman. New York and London: G. P. Putnam's Sons, 1912.

Committee of Fourteen. *The Social Evil in New York City: A Study of Law Enforcement by the Research Committee of the Committee of Fourteen*. New York: A. H. Kellogg, 1910.

Complete Bachelor Manners for Men. New York: D. Appleton, 1896.

Conwell, Russell. *Acres of Diamonds*. New York: Harper & Brothers, 1915.

Cook, Clarence. "Studio-Suggestions for Decoration." *Quarterly Illustrator* 4 (1895), 235–236.

———. *Art and Artists of Our Time*. 3 vols. 1888; rpt. New York: Garland, 1977.

Courtwright, David. *Dark Paradise: Opiate Addiction in America Before 1940*. Cambridge, Mass.: Harvard University Press, 1982.

Craib, Alexander. *America and the Americans*. London: Alexander Gardner, 1892.

Crane, Stephen. *Complete Short Stories*. Ed. Thomas A. Gullason. Garden City, N.Y.: Doubleday, 1963.

Crawford, F. Marion. *Dr. Claudius: A True Story*. New York: Macmillan, 1912.

Davis, Richard Harding. "The West and East Ends of London." *Harper's Monthly* 88:524 (December 1893), 279–292.

Davis, Tracy. *Actresses as Working Women: Their Social Identity in Victorian Culture*. London: Routledge, 1991.

Day, Samuel P. *Life and Society in America*. 2 vols. London: Newman, 1880.

DeBary, Richard. *The Land of Promise: An Account of the Material and Spiritual Unity of America*. London: Longmans, Green, 1908.

DeLisle, Winefred Mary. *Journal of a Tour in the United States, Canada and Mexico*. London: Sampson Low, Marston, 1897.

D'Emilio, John, and Estelle Freedman. *Intimate Matters: A History of Sexuality in America*. New York: Harper & Row, 1988.

Douglas, Mary. *The World of Goods: Towards an Anthropology of Consumption*. London: Allen Lane, 1979.

Du Maurier, George. *Trilby*. 1894; rpt. New York: Oxford University Press, 1995.

Elliott, Maud Howe. *This Was My Newport*. Cambridge, Mass.: Mythology Company, A. Marshall Jones, 1944.

Enstad, Nan. *Ladies of Labor, Girls of Adventure*. New York: Columbia University Press, 1999.

Erenberg, Lewis A. *Steppin' Out: New York Nightlife and the Transformation of American Culture 1890–1930*. Westport, Conn.: Greenwood Press, 1984.

Estournelles de Constant, Paul H. B. *America and Her Problems*. New York: Macmillan, 1915.

E. T. L. "Studio-Life in New York." *Art Journal*, 1877, 267–268.

Famous Pictures. Chicago: Thompson & Thomas, 1902.

Fiske, Stephen. *Off-hand Portraits of Prominent New Yorkers*. New York: George R. Lockwood & Son, 1884.

Fitz-Gerald, William [pseud. Ignatius Phayre]. *America's Day: Studies in Light and Shade*. London: Constable, 1918.

Fogelson, Robert M. *America's Armories: Architecture, Society, and Public Order*. Cambridge, Mass.: Harvard University Press, 1989.

Francis, Alexander. *Americans: An Impression*. New York: D. Appleton, 1909.

Fraser, James Nelson. *America Old and New: Impressions of Six Months in the States*. London: John Ouseley, 1912.

Freeman, Edward. *Some Impressions of the United States*. New York: Holt, 1883.

Freeman, James Edward. *Gatherings from an Artist's Portfolio*. New York: D. Appleton, 1883.

Furniss, Harry. *The Confessions of a Caricaturist.* 2 vols. New York: Harper & Brothers, 1902.

Gallati, Barbara Dayer. *William Merritt Chase.* New York: Abrams, 1995.

Gerdts, William H. *The Great American Nude.* New York: Praeger, 1974

Gere, Charlotte. *Nineteenth-Century Decoration: The Art of the Interior.* New York: Abrams, 1989.

Gerhard, William Paul. *American Practice of Gas Piping and Gas Lighting in Buildings.* New York: McGraw, 1908.

"Gérôme's Sword-Dance." *Art Journal,* 1877, 164.

Gilfoyle, Timothy J. *City of Eros.* New York: Norton, 1992.

Glazier, Willard. *Peculiarities of American Cities.* Philadelphia: Hubbard Brothers, 1885.

Godkin, E. L. "The Expenditure of Rich Men." *Scribner's Magazine* 20:4 (October 1896), 495–501.

Goffman, Erving. *Interaction Ritual: Essays in Face-to-Face Behavior.* Chicago: Aldine, 1967.

"Good Art as an Advertisement." *Architectural Record* 12:1 (May 1902), 111–115.

Gordon, Harry Panmure. *The Land of the Almighty Dollar.* London: Frederick Warne, 1892.

Gorky, Maxim. "Boredom." *Independent* 63 (8 August 1907), 312.

Griffin, Lepel Henry. *The Great Republic.* 1884; rpt. New York: Arno Press, 1974.

Gulick, Luther H. *The Efficient Life.* New York: Doubleday Page, 1907.

Hall, Bolton. *Three Acres and Liberty.* New York: Macmillan, 1907.

Hannay, James O. *See* Birmingham, George A.

Hardman, William. *A Trip to America.* London: T. Vickers Wood, 1884.

Hardy, Iza Duffus. *Between Two Oceans: or, Sketches of American Travel.* London: Hurst & Blackett, 1884.

Hardy, Mary McDowell Duffus. *Through Cities and Prairie Lands: Sketches of an American Tour.* 1881; rpt. New York: Arno Press, 1974.

Harrison, Frederic. "Impressions of America." *Nineteenth Century* 49 (1901), 913–930.

———. *Memories and Thoughts: Men-Books-Cities-Art.* London: Macmillan, 1906.

Hatton, Joseph. *To-day in America: Studies for the Old World and the New.* 2 vols. London: Chapman & Hall, 1881.

Hay, Mary Cecil. "The London Royal Academy." *Art Journal,* 1880, 218–221.

Hogan, James F. *The Australian in London and America.* London: Ward & Downey, 1889.

Hook, Philip, and Mark Poltimore. *Popular Nineteenth-Century Painting*. Suffolk, England: Antique Collectors' Club, 1986.

Howard, Joseph. *The Life of Henry Ward Beecher*. Philadelphia: Hubbard Brothers, 1887.

Howe, William F., and Abraham Hummel. *In Danger; or, Life in New York: A True History of a Great City's Wiles and Temptations*. New York: J. S. Ogilvie, 1888.

Howe, Winifred. *A History of the Metropolitan Museum of Art*. 2 vols. New York: Columbia University Press, 1913.

Howells, William Dean. *A Hazard of New Fortunes*. 1891; rpt. New York: E. P. Dutton, 1952.

Hudson, T. S. *A Scamper Through America, or, 15,000 Miles of Ocean and Continent in 60 days*. London: Griffith & Farran, 1882.

Hughes, Rupert. *The Real New York*. New York: Smart Set Publishing Company, 1904.

Hughes, Thomas. *Vacation Rambles*. New York: Macmillan, 1895.

Humphreys, Eliza M. *America—Through English Eyes*. London: Sidney Paul, 1910.

Ingersoll, Ernest. *A Week in New York*. New York: Rand McNally, 1891.

Irwin, Robert. *The Arabian Nights: A Companion*. London: Penguin, 1994.

James, Henry. *Daisy Miller*. 1878; rpt. Oxford and New York: Oxford University Press, 1985.

———. *The American Scene*. New York: Harper & Brothers, 1907.

Kabbani, Rana. *Europe's Myths of Orient*. Bloomington, Ind.: Indiana University Press, 1986.

Keating, John D. *The Lambent Flame*. Carlton, Victoria, Australia: Melbourne University Press, 1974.

Kendall, John. *American Memories: Recollections of a Hurried Run Through the United States During Late Spring of 1896*. Nottingham, England: W. Burrows, 1896.

Kipling, Rudyard. *American Notes*. Boston: Brown, 1899.

Kirkwood, John. *An Autumn Holiday in the United States and Canada*. Edinburgh, Scotland: Andrew Elliott, 1887.

Klein, Abbé Félix. *In the Land of the Strenuous Life*. Chicago: A. C. McClurg, 1905.

Klein, Maury. *The Life and Legend of Jay Gould*. Baltimore: Johns Hopkins University Press, 1986.

Kroupa, B. *An Artist's Tour: Gleanings and Impressions of Travels*. London: Ward & Downey, 1890.

Lamprecht, Karl. *Americana*. Freiburg, Germany: H. Heyfelder, 1906.

Leach, William. *Land of Desire: Merchants, Power, and the Rise of a New American Culture*. New York: Pantheon, 1993.

Lehr, Elizabeth Drexel. *"King Lehr" and the Gilded Age*. Philadelphia: J. B. Lippincott, 1935.

Leland, Lilian. *Traveling Alone: A Woman's Journey Around the World*. New York: American News, 1890.

Leng, John. *America in 1876: Pencillings During a Tour in the Centennial Year*. Dundee, Scotland: Dundee Advertiser Office, 1877.

Leslie, Shane. *American Wonderland: Memories of Four Tours in the USA, 1911–1935*. London: M. Joseph, 1936.

Lessard, Suzannah. *The Architect of Desire: Beauty and Danger in the Stanford White Family*. New York: Delta, 1996.

Lexow Committee. *Report and Proceedings of the Senate Committee Appointed to Investigate the Police Department of the City of New York*. 5 vols. Albany, N.Y.: James B. Lyon, State Printer, 1895.

Lind, Earl [pseud. Ralph Werther and Jennie June]. *The Female-Impersonators*. New York: Medico-Legal Journal, 1918.

Loti, Pierre. "Impressions of New York." *Century Magazine* 85 (1912–1913), 609–613, 758–762.

Lovett, Richard. *United States Pictures Drawn with Pen and Pencil*. London: Religious Tract Society, 1891.

Low, Alfred Maurice. *America at Home*. London: George Newnes, 1908.

Lowe, David. *Stanford White's New York*. New York: Doubleday, 1992.

Lubow, Arthur. *The Reporter Who Would Be King: A Biography of Richard Harding Davis*. New York: Scribner's, 1992.

Lucy, Henry William. *East by West: A Journey in the Recess*. 2 vols. London: Richard Bentley, 1885.

Mackenzie, John. *Orientalism: History, Theory and the Arts*. Manchester, N.Y.: Manchester University Press, 1995.

MacRae, David. *America Revisited*. Glasgow, Scotland: John Smith & Son, 1908.

Mahdi, Muhsin, ed. *The Arabian Nights*. Trans. Husain Haddawy. New York: Norton, 1990.

Marshall, Walter G. *Through America: or, Nine Months in the United States*. London: Sampson Low, Marston, Searle, & Rivington, 1881.

Martin, Frederick Townsend. *The Passing of the Idle Rich*. New York: Doubleday Page, 1912.

Masters in Art. Vol. 7. Boston: Bates & Guild, 1906.

Mather, Frank Jewett. *The Collectors*. New York: Holt, 1912.

Maurice, Arthur Bartlett. *Fifth Avenue*. New York: Dodd, Mead, 1918.

McAllister, Ward. *Society as I Have Found It*. New York: Cassell, 1890.

Meakin, Annette M. B. *What America Is Doing: Letters from the New World*. Edinburgh, Scotland: William Blackwood & Sons, 1911.

Miller, Daniel. *Acknowledging Consumption*. New York: Routledge, 1997.

Mizejewski, Linda. *Ziegfeld Girl: Image and Icon in Culture and Cinema*. Durham, N.C.: Duke University Press, 1999.

Money, Edward. *The Truth About America*. London: Sampson Low, Marston, Searle, & Rivington, 1886.

Montgomery, Maureen. *Gilded Prostitution: Status, Money and Transatlantic Marriages, 1870–1914*. London: Routledge, 1989.

Moran, John. "Studio-Life in New York." *Art Journal*, 1879, 343–345.

———. "Artist-Life in New York." *Art Journal*, 1880, 57–58, 121–124.

Morgan, H. Wayne. *New Muses: Art in American Culture 1865–1920*. Norman Okla.: University of Oklahoma Press, 1978.

———. *Drugs in America: A Social History, 1800–1980*. Syracuse, N.Y.: Syracuse University Press, 1981.

Muirhead, James F. *America: The Land of Contrasts. A Briton's View of His American Kin*. London: John Lane, Bodley Head, 1898.

Mulhall, William. "The Golden Age of Booze." *Valentine's Manual of Old New York*. New York: Valentine's Manual, 1923.

Munsterberg, Hugo. *The Americans*. Trans. Edwin B. Holt. New York: McClure, Phillips, 1904.

Myers, Denis Peter. *Gaslighting in America*. New York: Dover, 1990.

Nasaw, David. *Going Out: The Rise and Fall of Public Amusements*. New York: Basic Books, 1993.

Naylor, Robert Anderton. *Across the Atlantic*. Westminster, England: Roxburghe Press, 1893.

Nesbit, Evelyn. *The Story of My Life*. London: J. Long, 1914.

———. *Prodigal Days: The Untold Story*. New York: J. Messner, 1934.

O'Connor, Richard. *Courtroom Warrior: The Combative Career of William Travers Jerome*. Boston: Little, Brown, 1963.

Offenbach, Jacques. *Orpheus in America*. Trans. Lander McClintock. 1876; rpt. Bloomington Ind.: Indiana University Press, 1957.

Old and Modern Paintings, Water Colors, and Drawings Belonging to the Estate of the Late Stanford White, 1907.

O'Rell, Max. See Blouet, Paul.

"Oriental Embroideries." *Art Journal*, 1877, 132.

Pairpoint, Alfred. *Rambles in America, Past and Present*. Boston: Alfred Mudge & Son, 1891.

Parkhurst, Charles. *My Forty Years in New York.* New York: Macmillan, 1923.

Peiss, Kathy. *Cheap Amusements: Working Women and Leisure in Turn-of-the-Century New York.* Philadelphia: Temple University Press, 1986.

Peltre, Christine. *Orientalism in Art.* Trans. John Goodman. New York: Abbeville, 1998.

Penzel, Frederick. *Theater Lighting Before Electricity.* Middletown, Conn.: Wesleyan University Press, 1978.

Pettengill, Lillian. *Toilers of the Home: The Record of a College Woman's Experience as a Domestic Servant.* New York: Doubleday Page, 1905.

Pisano, Ronald G. *A Leading Spirit in America Art: William Merritt Chase.* Seattle: Henry Art Gallery, University of Washington, c. 1983.

Pocock, Roger. *Rottenness: A Study of America and England.* London: Neville Beeman, 1896.

Porteous, Archibald. *A Scamper Through Some Cities of America.* Glasgow, Scotland: D. Bryce & Son, 1890.

Rae, William Fraser. *Columbia and Canada: Notes on the Great Republic and the New Dominion.* London: Daldy, Isbister, 1877.

Rheims, Maurice. *The Strange Life of Objects.* New York: Atheneum, 1961.

Rosenberg, Charles. *No Other Gods: On Science and American Social Thought.* Baltimore: Johns Hopkins University Press, 1997.

Rosenzweig, Roy, and Elizabeth Blackmar. *The Park and the People: A History of Central Park.* Ithaca, N.Y.: Cornell University Press, 1992.

Roth, Leland M. *McKim, Mead and White, Architects.* New York: Harper & Row, 1983.

Rousiers, Paul de. *American Life.* Trans. A. J. Herbertson. Paris, France: Firmin-Didot, 1892.

Rovere, Richard H. *Howe & Hummel: Their True and Scandalous History.* New York: Farrar, Straus, 1947.

Sala, George. *America Revisited.* London: Vizetelly, 1883.

Santayana, George. *The Sense of Beauty: Being an Outline of Æsthetic Theory.* New York: Charles Scribner's Sons, 1896.

Scudder, Janet. *Modeling My Life.* New York: Harcourt Brace, 1925.

Shaw, Jennifer. "The Figure of Venus: Rhetoric of the Ideal and the Salon of 1863." *Art History* 14:4 (December 1991), 540–570.

Sheldon, George William. *Artistic Houses.* 1883; rpt. New York: Benjamin Blom, 1971.

Sherwood, M. E. W. *An Epistle to Posterity.* New York: Harper & Brothers, 1897.

Shinn, Earl [pseud. Edward Strahan]. *The Art Treasures of America.* 3 vols. Philadelphia: Gebbie & Barrie, 1879.

Simmons, Edward. *From Seven to Seventy*. New York: Harper & Brothers, 1922.

Smith, Samuel. *America Revisited*. Liverpool, England: Turner, Routledge, 1896.

Smith, T. *Rambling Recollections of a Trip to America*. Edinburgh, Scotland: D. S. Stewart, 1875.

Soissons, G. J. R. *A Parisian in America*. Boston: Estes & Lauriat, 1896.

St. Denis, Ruth. *An Unfinished Life: An Autobiography*. New York: Harper & Brothers, 1939.

Stead, William T. *Satan's Invisible World Displayed: or, Despairing Democracy: A Study of Greater New York*. London: Review of Reviews Annual, 1898.

———. *The Americanization of the World: or, the Trend of the Twentieth Century*. New York: Horace Markley, 1902.

Steevens, G. W. *The Land of the Dollar*. New York: Dodd, Mead, 1898.

Stewart, Robert. "Clubs and Club Life in New York." *Munsey's Magazine* 22 (October 1899), 106–122.

Stewart, Susan. *On Longing: Narratives of the Miniature, the Gigantic, the Souvenir, the Collection*. Baltimore: Johns Hopkins University Press, 1984.

Strange, Michael. *Who Tells Me True*. New York: Scribner's, 1940.

Sumichrast, Frederick de. *America and the Britons*. New York: D. Appleton, 1914.

Tales from the Thousand and One Nights. Trans. N. J. Dawood. London: Penguin, 1954.

Tarver, Edmund. "Artists' Studios." *Art Journal*, 1880, 273–276.

Thompson, Michael. *Rubbish Theory*. Oxford and New York: Oxford University Press, 1979.

Thomson, James. "Cozy Corners and Ingle Nooks." *Ladies' Home Journal* 10:2 (November 1893), 27.

Tod, John [pseud. John Strathesk]. *Bits About America*. Edinburgh, Scotland: Oliphant, Anderson, & Ferrier, 1887.

Vaile, P. A. *Y., America's Peril*. London: Francis Griffiths, 1909.

Van Dyke, John. *The New New York*. New York: Macmillan, 1909.

Vay de Vaya, Peter. *The Inner Life of the United States*. New York: E. P. Dutton, 1908.

Verrier, Michelle. *The Orientalists*. New York: Rizzoli, 1979.

Vincent, Ethel. *Forty Thousand Miles over Land and Water*. 2 vols. London: Sampson Low, Marston, Searle, & Rivington, 1885.

Vivian, Henry Hussey. *Notes of a Tour in America*. London: Edward Stanford, 1878.

Wagner, Charles. *My Impressions of America*. 1906; rpt. New York: Arno Press, 1974.

Wall, Evander Berry. *Neither Pest Nor Puritan*. New York: Dial Press, 1940.

Warner, Charles Dudley. *Fashions in Literature*. New York: Dodd, Mead, 1902.

Warner, Charles Dudley. *The Golden House*. New York: Harper & Brothers, 1902.

Welling, Richard W. G. *As the Twig Is Bent*. New York: G. P. Putnam's Sons, 1942.

Wells, H. G. *The Future in America: A Search After Realities*. New York: Harper & Brothers, 1906.

Wells, Richard A. *Manners, Culture and Dress of the Best American Society*. Springfield, Mass.: King, Richardson, 1890.

Wharton, Edith. *A Backward Glance*. New York: D. Appleton-Century, 1934.

———. *The Age of Innocence*. New York: Modern Library, 1948.

Wheeler, Candace. *Yesterdays in a Busy Life*. New York: Harper & Brothers, 1918.

Whibley, Charles. *American Sketches*. Edinburgh, Scotland: William Blackwood & Sons, 1908.

Wiggin, Kate Douglass, ed. *The Arabian Nights*. New York: Charles Scribner's Sons, 1909.

Woodley, William. *Impressions of an Englishman in America*. New York: W. J. Woodley, 1910.

Wright, Mabel Osgood. *My New York*. New York: Macmillan, 1926.

Wu, T'ing-fang. *America Through the Spectacles of an Oriental Diplomat*. 1914; rpt. Taipei, Taiwan: Ch'Eng-Wen Publishing Company, 1968.

INDEX